W9-BDT-151

DARK SKIES

A Practical Guide to Astrotourism

Contents

© Gianni Triggiani / 500 px; Previous spread: © RGB Ventures / Alamy Stock Photo

Foreword

by Phil Plait (aka the Bad Astronomer)

We had been driving for hours, my wife and I, and were exhausted when we finally arrived at the dude ranch in western Colorado. The drive through the mountains had been gorgeous but extended, and the sun was going down when we finally pulled in.

We ate dinner by twilight, and before we could even really unpack our bags, the light in the sky was gone and the view out our cabin window was utterly black. I suggested sleep, but Marcella had different plans. 'I want to see how the sky looks', she suggested, and how could I resist? I found our coats and red-covered flashlights, and we faced the brisk (and thin) air to start walking along the creek that flowed behind the cabin.

We picked our way along a trail, sleepiness and unfamiliarity slowing our progress. Rural Colorado has plenty of wildlife, and I kept a wary eye ahead or to the sides, peering into the darkness of the trees around us to make sure we were the only ones enjoying the outdoors.

After a few hundred yards the trail opened up into a clearing. I was still wearily scanning around us for critters when I heard my wife gasp. I turned and saw that she had turned off her flashlight and was staring up, mouth open, gawking. Clicking my flashlight off as well, I followed her gaze upward, already knowing what I'd see. And I was right.

Thousands of stars shone down on us. The Milky Way, usually a barely hinted-at phantom from our home, was like a solid structure arcing over our heads. So many faint stars were visible that it took me a moment to get my bearings, even though I've been an astronomer my whole life and have spent thousands of hours under the night sky.

Myriad thoughts jammed into my head, expressions of awe and wonder and beauty. But Marcella was

able to synthesise all of them into a single breathy, exhaled sigh. 'Oh my God,' she muttered. I smiled, agreeing. Oh, how I missed this view!

The thing is, I'm lucky. *I knew what I was missing.*

In 2016 a study published in *Science Advances* mapped the glow of the sky caused by artificial lights and compared it to population location. The result was jolting: about 80% of the world's population lives under light-polluted skies. In the US that number jumps to 99%. Hundreds of millions of people in the US cannot go outside, look up and see more than a handful of stars. We have erased the treasures of the universe from our lives, that light that had travelled for centuries, millennia, to reach our eyes.

Of course illumination at night is important. We humans aren't all that good at being tied to the sun; our civilisation's activity has spilled over into the night, and we need to see what we're doing. This in itself isn't the issue; the problem is light that's wasted, shining up instead of down, thrown into the sky, illuminating the air around us and washing out the stars. Millions of people grow up under a sky that, at night, is as lifeless and dull as a pothole filled with brackish water.

Not to be too flatly obvious about it, but *the entire universe is up there*, and we're missing it. Even in our own atmosphere, there are wonders to see. Shooting stars, meteors created from tiny bits of interplanetary flotsam shed from passing comets, burning up 50 miles (80km) above our heads as they ram through the air at hypersonic speeds. Aurora flowing and glimmering, caused by the wind from the sun interacting with molecules in the air even higher up, creating cascades of electrons jumping around their atomic hosts, emitting individual photons of light in numbers too high to count.

Continuing on past our atmosphere, there are satellites to watch as they orbit in near-earth space; the planets as they circle the sun; stars and clusters and nebulae; and galaxies that are so numerous that a grain of sand held at arm's length blocks out tens of thousands of them.

This is what we're missing. But there's some good news in all this: it doesn't have to be this way.

Dark-sky advocates are gaining traction, able to convince authorities that we don't necessarily need fewer lights, we need *smarter* ones – lamps that illuminate the ground and not the heavens. Designated dark-sky sites are growing in number and popularity, and more people are becoming aware of astrotourism and heading out into less-populated areas to see the skies.

Even better, you need not go too far to sky gaze. Sometimes the nearest window will suffice. You can see lunar eclipses from anywhere, as the shadow of the Earth slowly covers the moon, and solar eclipses are of necessity observable during the day. The planets are visible even from the downtowns of major cities, and the moon is always a joy to see even with modest optical equipment.

Still, there's so much more to see, and places to go to see it. And here's the best part yet: in the pages of this book you have a guide in Valerie Stimac. She shows you how to find your way out and up to the best places to experience all the different wonders the sky has in wait for you. Heed her advice, and the whole of the cosmos is open to you.

Standing in that clearing in the remote Colorado Rockies with my wife that night, the two of us slack-jawed and silent, soaking in the constellations, a familiar thought crossed my mind. I have always been amused at the expression we use when we want to leave our crowded, overlit cities and suburbs and head out to some remote, quiet, dark location: 'getting away from it all.'

That's *exactly* backward. When you travel away from those lights, you leave behind nothing: a blank sky. When you travel to these dark sites, you fill that blankness with the entire universe above you. You're not getting away from it all. You're heading towards it.

© Wanson Luk / 500 px

Introduction
by Valerie Stimac

When the cloudy expanse of the Milky Way stretches above us from horizon to horizon, or a meteor streaks across the sky, or a rocket defies gravity to leave Earth, it touches on a sense of wonder and awe. There is something breathtaking and humbling about the knowledge that beyond the protective layer of our atmosphere, there is a lot more out there. The universe is vast almost beyond comprehension: while technology helps increase our knowledge of moons, planets, and suns, we can hardly imagine how many other places there are in our solar system, galaxy, and the universe once you leave planet Earth.

The natural world on Earth never ceases to amaze us; we make pilgrimages to Everest, Niagara, the Amazon, and countless other awe-inspiring sites on our bucket list. But somehow, the night sky is often omitted from the list of natural experiences we should seek out. Yet its magnificence can be even more overwhelming than terrestrial wonders. For millions of years, the stars have wheeled overhead, and the planets have performed their celestial dance. Observing this pageant used to be a nightly ritual for humans across the planet until very recently. But while we often book trips to explore new cities and try new foods, we rarely do the same for astronomical phenomena and space experiences. We may have gone stargazing as a kid or learned about astronomy in school, but we don't journey to discover it firsthand. In not seeking out encounters with astronomical phenomena, whether at the certified dark sky parks listed in this book or by viewing a meteor shower or eclipse, we deprive ourselves of a magical experience. Less than one hundred years ago, seeing the unobscured night sky was a birthright; now it is inaccessible to urban and suburban residents across the globe. Yet it's still in reach if we seek it out.

What drives our collective interest in the night sky? It's likely the case that the roots of astronomy lay deep in the prehistoric era, among the first Homo sapiens who became aware that the movements of the sun, moon, and stars were not random. In an attempt to find significance among these patterns, religious beliefs were established to help make sense of the natural phenomena. These religious beliefs remain closely tied to astronomy to this day, as reflected by the practice of astrology (the idea that the movement and placement of stars and planets have a direct impact on our daily lives).

A more modern interpretation might also say that though we did not always know with scientific certainty that there was 'more' beyond Earth, our human nature to explore and colonise drives us to look toward the stars. In the 21st century, it's likely that our efforts and investment will take us to other planets in our solar system at the very least. While we have spent centuries learning about the night sky, our time exploring it has only just begun.

The most easily accessible way to enjoy the night sky is by stargazing, looking up at the constellations and planets visible either with the naked eye or through a telescope. Astronomy dates back nearly 5000 years, to the Bronze Age. While there aren't many records from this time, archaeoastronomers have discovered evidence and relics that suggest that from among the earliest periods in human history, we were attentive to the night sky and attempted to record the patterns observed there. Nearly every major civilisation at one time was involved in the study of astronomy. Major sites testifying to the astronomical knowledge of earlier cultures remain in the Yucatan, at Uxmal and Chichen Itza; at Chaco Canyon (p94); and at the pyramids of Egypt, Stonehenge, and more.

Early contributions by civilisations like the Sumerians, Babylonians, and Indians are still used in astronomy today. Over the centuries, the amalgamation of work by Chinese, Islamic, Egyptian, and European astronomers helped solidify astronomy into a scientific field in its own right. During the medieval era, astronomy was advanced significantly by the work of Islamic astronomers. While astronomy was actively practiced in Asia, Islamic astronomers helped with the translation from ancient Greek to Latin of fundamental astronomy texts by thinkers including Aristotle, Euclid and Ptolemy. As a result European astronomers were able to recommit to the science of astronomy that was at risk of being lost. Islamic astronomers also created some of the most accurate calendars, predictive models, and recorded observations of astronomical phenomena in human history.

Later, Renaissance astronomer Nicolaus Copernicus helped initiate the Scientific Revolution which fundamentally shifted human understanding of astronomy and science in the 15th and 16th centuries. While the idea of a heliocentric universe had been proposed centuries before by the Greek astronomer Ptolemy, Copernicus' reassertion that the Earth orbited the sun became one of the most controversial ideas of human history. Physicists and astronomers including Galileo, Kepler, and Newton helped drive forward our understanding of the universe using this new model.

After initial resistance by the Catholic Church, the Copernican theory was accepted and knowledge about astronomy and astrophysical principles began to receive global consensus. The Copernican revolution came to its natural conclusion with the discovery of a series of scientific laws that helped us understand the night sky and our place in it. Between the 17th and 20th centuries, the rate of discoveries by observational astronomers increased exponentially as the astronomical objects and phenomena laid out by these new laws was confirmed. Findings included the discoveries of planets, moons, asteroids, and comets in our solar system as well as more distant galaxies, nebulae, exoplanets, and black holes. New observing technology developments accelerated our rate of discoveries too. First invented in 1608, simple telescopes became increasingly powerful at observing the heavens. While there continued to be disagreement in reconciling religious and scientific beliefs about the solar system and universe, these were for the most part relegated away from the steady advancement of human knowledge of planets and moons, asteroids and comets, nebulae, supernovas, and galaxies.

During the 20th century, massive strides were made to improve astronomical technology for observing the skies, and our theoretical understanding made similar strides after Einstein's breakthroughs around General and Special Relativity. The light we see from the stars and planets has to travel across space to reach Earth; as a result, understanding how light moves is fundamental to understanding astronomy. Building on the massive legacy from civilisations and centuries of astronomy, scientists have been able to ask the deepest questions about the origins of the universe – and have begun to craft answers based on the observational data and theoretical models we have developed. As the 21st century continues, we are closer than ever to understanding the night sky, but still have a lot we don't even know to ask. When we look up at the stars and galaxies in the night sky, we are seeing the death of old stars in supernova, the birth of new ones in stellar nebulae (also sometimes called 'nurseries'), and in some cases, the impacts of invisible-to-us black holes on the space around them.

Even the space race of the 20th century was in part driven by the pull we feel to reach the stars. Milestones such as putting the first man in space or reaching the moon were meaningful because they took significant steps beyond our planet – the place we have called home for millennia. The human race to space has continued to launch satellites, orbiters, space telescopes, and rovers to explore deeper into space to better understand how the universe works.

The latter half of the 20th century and early 21st century has also seen the rise of the dark-sky preservation movement. Driven by international organisations like the International Dark-Sky Association, national bodies like the Royal Astronomical Society of Canada, and local institutions and advocacy groups, there is an increasing focus on preserving the dark sky where it is still visible, or in some cases reverting back to darker skies through infrastructure planning and lighting replacements. Many of the locations mentioned in this book are destinations focused on dark-sky preservation, and some have received designations for their work preserving the darkness. If you think you've seen the night sky but you've never witnessed it from a location with a truly dark sky, you're in for the surprise of a lifetime.

The skies above us are part of our heritage, both natural and cultural. Astronomy and stargazing are an important part of human history, one that can connect us back to early myth or awaken us to the vast scale of our universe and its many mysteries. Witnessing the sweep of the Milky Way, the remains of passing comets as they burn up in our atmosphere, or the shimmering aurora, we better understand space and our place in it. This book will help you experience all of this and more first-hand, so that you can glimpse some of the celestial wonder yourself. Taking time to enjoy the heavens, whether on travels abroad or from the backyard, deepens our knowledge and appreciation for our planet and the universe as a whole.

How to Use This Book

This book is divided into sections based on whether it's about witnessing the dark sky at a specific preserve, observing a natural phenomena such as a meteor shower, eclipse, or the aurora up close, or travelling to a major telescope or laboratory. You can even explore the options for suborbital space flight! Alternatively, maybe you won't be travelling anywhere but up, by looking at the heavens directly in your backyard. In each section, you'll learn about different space-related activities, then gain an understanding of where and how to have that experience.

Stargazing focuses on the basics of appreciating the dark sky. In it, you'll find an overview of how to stargaze and what types of objects you can look for in the night sky. You'll also find tips on urban stargazing, which most people can do no matter where they live. There's information on getting more involved with stargazing communities, including by joining astronomy clubs, visiting observatories, and attending star parties. You'll get some tips on how to shoot astrophotography and how you can give back through citizen science that uses the support of ordinary people on Earth to analyse and answer some of our biggest questions about space. This is your primer on how to demystify the sky.

Dark Places is devoted to highlighting 35 of the best places around the globe for stargazing and experiencing the night sky. While this section is far from exhaustive, it includes locations across the globe that are designated dark-sky places. These designated places take additional measures to reduce light pollution and ensure that if the skies are clear you'll see the stars (nothing can help you overcome a cloudy night, alas). There are also dark places in this section which do not have a formal designation but which possess a special attraction for stargazing and astrotourism.

Astronomy in Action focuses on destinations and experiences where you can get a closer look at space science. In this section you'll recognise some of the world's top research facilities and observatories. Most of these locations are open to the public (though often on limited schedules) meaning you can plan to visit them on a trip to the area – or even make astronomy the sole focus of your trip.

Meteor Showers has everything you need to know about some of the most consistent and impressive meteor showers throughout the year. Meteor showers occur throughout the year on a regular schedule. You'll learn about the science behind meteor showers, when they occur and nights of peak activity, and where to look in the night sky to try and see meteors.

Aurora is devoted to another breathtaking astronomical phenomena well worth a special trip: the stunning aurora. The section is subdivided into two parts, focusing on the aurora borealis in the northern hemisphere and the aurora australis in the southern hemisphere. You'll get guidance on when and where to see the aurora in each country where they commonly occur, plus some tips on other destinations where it's possible (but rare) to see the aurora during particularly strong magnetic storms.

Eclipses is devoted to the science and schedule of total solar eclipses in the next decade. If you've never experienced totality, this is a starting guide on how to plan your trip and become an eclipse chaser. You'll learn about where the path of totality will occur for each eclipse, plus how to get there and what else you can experience in the region once the solar eclipse is over.

Launches helps you experience a different side of astrotourism: rocket launches. Countries around the world are actively launching rockets to send instruments, supplies, and humans to space, and you can travel to launch locations around the globe.

Space Tourism discusses the future of humans in space – including you! In this section you'll learn about the major players in the rapidly evolving space tourism market, plus some of the common destinations and experiences for going to space.

Whether you visit a professional observatory, take part in Space Camp, observe a meteor shower, or merely spot the constellations above, the options are endless, and they'll reveal a world beyond imagining. This book is not an encyclopedia of astronomy or the comprehensive guide to all space experiences in the world. There are many places not listed here where you can enjoy the aurora, see meteors and stars in a dark night sky, and even marvel at the human ingenuity that is making us a multi-planetary species. Instead, use this book as inspiration. Let it spark the idea that you can enjoy your next destination after the sun goes down, add on a few extra days for one of these experiences, or plan a trip to enjoy the night sky anywhere in the world.

Stargazing

INTRODUCING the STARS

Seeing the night sky is one of the great wonders of the world, and it's among the easiest to experience. But when you first start stargazing at your dark-sky destination, it might be hard to know what you're seeing at first.

The sky is full of interesting objects, from single stars and planets to constellations and even galaxy clusters. The key is becoming familiar with the night sky and knowing where to look. Some of the basic things to hunt for in the night sky are the single stars and constellations that help orient us in the night sky. Constellations are groupings of stars that are near one another visually from our perspective on Earth, though they may be millions of light years apart in space. These stars create pictures in the sky that our ancestors used to explain natural phenomena, construct legends and myths, and pass down stories from human history.

Today, constellations help divide up the sky into areas we can observe and study. There are even different ones in the northern and southern hemispheres. All these objects move across the night sky throughout each month, year, and century on their own celestial journeys, and are visible across the globe at different times. In the northern hemisphere, the primary orienting stars are Polaris, the North Star, and Sirius, the brightest star in the sky (the only brighter objects are our moon and some of the visible planets like Venus, Jupiter, and sometimes Mars). Sirius is also visible from the southern hemisphere, as is Alpha Centauri, a multi-star system that's our closest celestial neighbour at 'just' four light years away.

The night sky is presently divided up into 88 constellations that have been universally (or should we say globally) agreed upon. This modern list was created in 1922 by organising and simplifying hundreds of constellations that had been developed and named over centuries of human history, often in conflict with one another. Some of these constellations may be familiar, such as Ursa Major (the Bear, inclusive of the Plough/Big Dipper), Orion (the Hunter), and Cygnus (the Swan or the Northern Cross). Others are more difficult to spot and more obscure in meaning to modern stargazers, though modern apps can demystify them. By identifying some of the most common constellations in your night sky during each season of the year, you can grow more familiar with the sky and begin to explore some of the deeper wonders in the universe beyond. Learning to identify the visible planets is another great way to connect with the tapestry above, as you may seek out Venus in its guise as morning or evening star or spy bright Jupiter.

© Dick Tang / 500 px

How to Stargaze

Many people learn to stargaze as a child when their parents or teachers point out some of the most famous stars and constellations in the night sky. While the names of these stars and constellations might vary depending on the culture or time you grew up in, they're a good foundation for stargazing later in life. If you didn't grow up with an opportunity to stargaze – either because it wasn't taught or because you lived in a city where stargazing wasn't possible – now's the time to begin. Stargazing is one of the easiest ways to enjoy nature here on Earth: all you need are your eyes and dark skies!

Before you go stargazing you may want to do some research on the constellations visible in the sky above you, depending on your location and the month. The good thing is that the movement of celestial bodies in space is predictable, and there are powerful tools to show you what the night sky will look like before you start stargazing (or historically, if you're interested in a significant day in the past). You can consult star charts or use a stellarscope to research what the sky will look like. A stellarscope looks like a small telescope but you can place a night sky map disk inside the tube and choose different settings on the outside to view the night sky through the viewfinder before the sun goes down. Several websites also offer tools where you can plug in your location, the date and the time you plan to go stargazing to view a simulation of the night sky at that time and place. In-the-Sky.org is one of the most powerful tools for doing this research. Plenty of apps offer a similar service too.

In addition to waiting for a night sky that is clear and free of clouds, it can also help to plan your stargazing sessions during a time each month when the moon is not full. The moon is the brightest object in the night sky, and it actually impedes our ability to see the stars and other celestial objects near it when it's fully bright. Plan to stargaze in the 20 days surrounding the new moon for the darkest skies and to see the greatest number of stars.

Once the sun goes down and you're ready to begin stargazing, you need to find a place where the skies are dark enough to see the stars. In most 21st-century cities this isn't possible, so you may need to travel outside of the city environs to one of the dark sky places mentioned in this book or do some research online about stargazing spots near your home. If you live outside the city, you may be able to stargaze from outside your home by turning off the lights and letting your eyes adjust. In either case, give your eyes up to 20 minutes to adjust to the darkness, and don't look at any lights or devices during that time. Once your eyes are fully adjusted, you'll see far more stars than you realised at first.

When you begin stargazing, there are a number of objects you can look for. In the northern hemisphere, Polaris (the North Star) is the star around which all other stars appear to rotate during year – it is located due north or above you in a northerly direction depending on the latitude you're stargazing from. Sirius is the brightest star in the night sky, and a good one to look for as you get oriented. Located in the constellation Canis Major (the Great Dog), Sirius reaches its highest position in the sky with each new year. In the southern hemisphere, Alpha Centauri is one of the brightest stars in the sky, located in the constellation Centaurus (the Centaur).

You can also look for other planets in our solar system while stargazing. Five planets are visible with the naked eye, no telescope needed. Mercury is the least-observed visible planet as it is typically close to the sun in the sky, but if you time your stargazing during the right month and time of day, it's possible to spot Mercury

GEAR YOU NEED FOR STARGAZING:

Warm clothes – after the sun goes down, the temperature often drops and extra layers will help keep you warm

Blanket/ground cloth – the easiest way to view the night sky is by lying down on your back; if you stand and look up, you may injure your neck

Red lights – if you need lighting while setting up or stargazing, make sure it is red, or cover your flashlights with red plastic wrap; our eyes dark sensitivity is less affected by red light, meaning it doesn't impede viewing

(Optional) Telescope or binoculars – if you want to look at objects that we can't see with our naked eyes, you'll need a tool that provides extra magnification. Alternatively, attending a star party gives you free access to the scopes of community members

near the horizon after the sun has set. Venus is the brightest planet in the sky, and the second brightest object in the night sky (after the moon). Venus' distinctive white-yellow hue makes it easy to spot. Mars is also visible from Earth, as it is our nearest planetary neighbour. Mars typically appears as an orange dot in the sky, making it distinctive and easy to spot. As the biggest planet, Jupiter appears quite bright in the night sky, and even Saturn is identifiable once you know what to look for. Unfortunately, Neptune and Uranus are not visible without binoculars or a telescope, but with the help of either, they too are findable and awe-inspiring.

There are other objects you can see without aid in the night sky. The most important is the Milky Way, our own galaxy visible as a cloudy band stretching across the sky. This visual 'cloudiness' is actually caused by the millions of stars and solar systems that make up our galaxy. From our position in a distant arm of our home galaxy, we can see a large portion of our celestial neighbourhood during different times of the year as our orbit and position allow us a glimpse of the many star systems that make up the Milky Way. For millennia, our ancestors were able to see the Milky Way almost every night, but urbanisation and light pollution have made this harder for most people who live in developed areas. This makes it even more important to travel out of urban areas and get some perspective on our size and place in the universe.

As the biggest planet, Jupiter appears quite bright in the night sky, and even smaller and fainter Saturn can be identifiable once you know what to look for.

We can also see other galaxies in the night sky. Some of our nearest intergalactic neighbours, the Large and Small Magellanic Clouds (or LMC and SMC), are visible in the southern hemisphere as two small cloudy areas in the sky. In the days of celestial navigation, they were known to the Māori, the Polynesians and the Khoisan culture of South Africa. In ancient Sri Lanka, what we know now to be faraway galaxies are thought to have been called the Mahameru Paruwathaya or "the great mountain", as they look like the peaks of a distant mountain range. Later they would be catalogued by Muslim astronomers as early as the 9th century.

The Andromeda Galaxy is also visible from across the Earth on moonless nights. At 2.5 million light years away and almost 10 times the width of the LMC, it is the nearest large galaxy to us. You can also see meteor showers and comets passing Earth, or witness eclipses; typically these astronomical events are widely publicised, such as when the Hale-Bopp comet passed Earth in 1997.

It's also possible to see man-made objects as they orbit Earth. One is the International Space Station (ISS), which is constantly orbiting 248 miles (400km) above Earth. Because it's so close, the ISS can be spotted during daylight as well as at night, passing above as a steady, bright object.

Getting Involved: Star Parties and Observatories

If you're interested in getting more involved in astronomy, one of the best ways is by joining your local astronomy club. Most people don't even know that there is an astronomy club in their area, but there are amateur astronomers all over the world and they occasionally meet up to share insights about space and do viewings together. Most astronomy clubs hold educational events such as lectures, talks, or networking events, observing events where members bring their telescopes to observe together, and public outreach events to educate the wider community about astronomy. There may be a small annual fee to be a member of your

© Stocktrek Images, Inc. / Alamy Stock Photo

local astronomy club, but this typically goes to support these events and help support astronomy research.

Another way to learn more about deep-space objects we can't see with our eyes – or to see these objects much more closely – is by visiting a local observatory. Observatories are scientific or community facilities that operate one or more telescopes to study the night sky and educate the public about it. Almost all observatories (including privately owned ones) typically have open nights for the public to come and view the sky through the equipment with the help of an astronomer or volunteer.

There are observatories located around the globe, often in dark-sky places away from major city lights. You may occasionally find an observatory in a city; typically the observatory was there long before the city grew up around it! While you might wonder if it's worth travelling to an observatory, most observatories have telescopes that are significantly more powerful than any telescope you can purchase on your own. There's no better way to view some celestial objects than through a telescope at an observatory. If you live in a city and can't travel to a dark-sky site, then your local planetarium is a great alternative. Ones like the Rose Center for Earth and Space at the American Museum of Natural History can be worth visiting in their own right, despite being located in

There's no better way to view many celestial objects than through a telescope at an observatory.

© M.Quinn / NPS

light-polluted cities.

Some observatories also hold star parties in partnership with a local astronomy club. During these events local astronomers bring their own telescopes, and everyone sets up to look at different objects in the sky. This gives you the chance to view a large number of fascinating night sky objects you might not otherwise see with the naked eye, like distant galaxies, nebulae and certain planets.

If you aren't located near an observatory or planetarium and can't travel to a dark-sky location to enjoy the night sky, it is still possible to do urban stargazing. If you're prepared not to see as many stars as a result of light pollution, it's still a fun activity to enjoy on a clear night. Since light pollution will interfere with your ability to view all but the brightest objects in the sky, get familiar with the visible planets (like Mars and Saturn), and pay attention to the phase of the moon when planning your stargazing. A full moon can cause light pollution of its own, making it even harder to spot constellations, planets or meteors in an already light-polluted area.

You need to find a location that has some protection from light. While you may not be able to reduce the light directly above, be sure to move away from or find a place which is shielded from light shining directly on you while you're stargazing. Public parks are a great option since they often have organised events with staff available.

© Liu Chaoyang / Getty Images

Astrophotography for Beginners

Sometimes, it's not enough to simply see the wonders of the night sky with your eyes or through a telescope. You may find that you want to learn how to capture photographs of the night sky, a branch of photography called astrophotography. While astrophotography may sound like a simple activity, it's actually a highly technical branch of photography that requires patience, the right equipment, and a willingness to learn.

Before you start trying to shoot any astrophotographs, ensure your camera is the same temperature as the air outside. If your camera is warmer than the ambient air temperature, the lens or sensor may fog up, and you'll end up with clouds in your photos that aren't the Milky Way or Magellanic clouds, but technical interference. If shooting in freezing temperatures, electronics may even shut down!

You'll also need certain specialised equipment to shoot astrophotography. In particular you will need a tripod and a timer or remote. One of the most critical parts of composing a good astrophotograph is ensuring that the photograph is stable, so the stars and other celestial objects will appear crisp and clear. The only way to do this is by using a tripod and timer or remote; if you use your hand to control the shutter button, your camera will move too much and your pictures will

EQUIPMENT BASICS

Here is a quick list of what you need to shoot stunning astrophotographs:

1 A camera with manual settings control. It's necessary to control three main settings (shutter speed, aperture, and ISO) on your camera for successful night photography.

2 A tripod and timer or remote to stabilise your photos.

3 Extra batteries, as cold night photography will drain batteries faster.

4 A solar filter for your lens if you're shooting specific astronomical phenomena like eclipses or solar flares.

end up blurry. Set up your tripod, affix your camera, determine your settings, and use the remote or timer to control the shutter. If it's windy, you may also need to stabilise your tripod so it doesn't move.

Another piece of equipment you'll need for astrophotography is extra batteries. As it's cooler at night, your camera batteries will drain faster. Additionally, long shutter speeds and your camera's photo processing will eat up extra energy compared to shooting photos during a warm day. Bring 2-3 extra batteries when you go out to shoot astrophotos so you'll have enough power to get through the whole session.

It's always important to consider composition when you're shooting your astrophotographs. While the stars might be your subject, it can help to add an object in the foreground for context. Some photographers use natural settings like mountains, trees, or rock formations to create a sense of interest beyond the stars themselves.

In terms of camera settings, astrophotography is a balancing act. For those familiar with photography, the trade-off comes between capturing enough light to see the stars, planets, aurora, or other objects you're photographing and having too much 'noise' in your photograph. There are three basic settings you control as part of astrophotography: your shutter speed, your f-stop, and your ISO. By finding the right balance of these three settings, you can capture the night sky in different ways.

When it comes to shutter speed, it might seem like longer is always better, but this depends on your other settings and what you're trying to capture. If you want to photograph star trails or the aurora, a longer shutter speed will be better. If you want crisp, clear stars, you'll need a shorter shutter speed but will have to adjust your other settings to let in enough light.

You can set your aperture to determine how much light you want to let in to your sensor. A lower aperture allows more light; but you'll need to balance this against your ISO (sensitivity of your sensor) and shutter speed to let in the right amount.

Your ISO (sensitivity) is the final piece of the puzzle. If you set your ISO too high, you'll end up with noise in your photograph – the colours will look fuzzy like a badly tuned TV. In general, you'll need a moderately high ISO (800-3200) to capture enough light but not have too much noise. Some photographers use a high ISO (3200+) and clean up the 'noise' in post-production and editing. Setting your focus is also an important step. While it might seem intuitive to set your focus as far as possible (infinity), most astrophotographers actually set focus a bit back from infinity to help make the photo as crisp and clear as possible. But the most important thing? Just to practise and learn by doing.

Citizen Science

Have you already conquered observing and dabbled in astrophotography, or do you simply prefer to stay in the warmth of your own home while pondering the majesty of the cosmos? Professional astronomers with PhDs and access to high-end observatories are a recent development; earlier astronomers were often amateurs and hobbyists.

Even today anyone familiar with the night sky can be the first to spot the appearance of a brightly shining supernova in the sky – and even confirm existing models of the early stages of a supernova – as happened with Argentinian amateur observer Victor Buso in September 2016. Buso and a fellow amateur observer would go on to be credited as co-authors of a paper in ***Nature*** about the discovery. Collaborations between professional astronomers and citizen scientists are a major asset in a field where the amount of data being captured by satellites and requiring review is so vast, and where even major ground-based observatories are limited in how much of the sky they can take in at once.

NASA is one of the biggest advocates of citizen scientist involvement and has created a variety of projects that can involve ordinary citizens and amateur astronomers. HubbleSite focuses specifically on observations made by the Hubble telescope while it was in service orbiting 340 miles (547km) above Earth. This space telescope produced some of the most magnificent space photos ever taken, and for each one you see in the news, there are tens of thousands more photos in the Hubble database. Occasionally NASA and Hubble scientists will come up with a new project, create a website, and put out a call for volunteers to help sift through the huge number of photo files to assist them in answering important questions. Past projects included helping determine the age of stars in the Southern Pinwheel Galaxy and providing insight into the movement of the Milky Way and Andromeda galaxies. When the Hubble telescope is replaced by the James Webb Space Telescope in the 2020s, it's likely NASA will set up a new website for citizen scientists to get involved.

Another major project that NASA supports is called Planet Hunters. Using data from the Kepler spacecraft to identify changes in light from distant stars, you can help researchers spot stars that may have planets orbiting them. The website has a course that teaches you about the light data from Kepler, plus how to spot a transit, when a planet moves in front of a star, temporarily dimming it. Once you get the hang of it, you can spend time reviewing data sets and flagging potential transits. Your work goes directly to researchers who review and approve any data that looks like a probable planetary system. The Kepler spacecraft is being replaced by the TESS (Transiting Exoplanet Survey Satellite) spacecraft, which will provide scientists with new data they'll need help sifting through. Yes, you really can discover new planets!

NASA also regularly publishes articles and stories that call for interested citizens to get involved. Typically these are shared on the website for specific missions, such

Anyone familiar with the night sky can be the first to spot the appearance of a brightly shining supernova . . . as happened with Argentinian amateur observer Victor Buso in September 2016.

as the Juno mission to Jupiter, which allowed amateur astrophotographers to share their photos of the planet. You can also look at popular astronomy publications like ***Sky & Telescope*** (www.skyandtelescope.com), which has a whole section of its website devoted to ways amateur and professional astronomers can work together.

One of the most well-known citizen science projects is part of SETI, the Search for Extraterrestrial Intelligence. Used broadly, SETI comprises a variety of research projects and companies focused at least in part on the search for life among the stars, and there are several community outreach projects. The most popular is called SETI@home, and anyone with a computer can participate. After you install the software on your computer, SETI@home uses spare processing power to help search through radio telescope data (including from Arecibo telescope in Puerto Rico, p120) for anomalies that might suggest alien life. Similar projects include MilkyWay@home, which is creating 3D models of the area around the Milky Way, and Einstein@Home, which searches for gravitational waves. As these programmes all run on your computer (typically when you're not using it), they are among the easiest ways to get more involved in the astronomical field as a citizen scientist. These are great for kids and students with an interest in science as well as hobbyists who want to dabble in astronomy outside of their day job or in retirement. Two of the other biggest citizen science platforms are Cosmoquest (https://cosmoquest.org/x) and Zooniverse (www.zooniverse.org).

Major astronomical events like the occurrence of a solar eclipse are another great chance to get involved as a citizen scientist. For example, during and after the 2017 total solar eclipse, a variety of projects sprang up that helped scientists better understand the eclipse and the behaviour of the sun's corona. Photographers were encouraged to submit their own photos and help sort through massive photo sets to help professional astronomers spot interesting phenomena. If you know a major event such as a solar eclipse is happening and you plan to witness it, reach out to your local astronomy club or national astronomical society (like NASA or the Royal Astronomical Society) to get involved.

Finally, you can get involved as a citizen scientist in astronomy with just a telescope, whether using your own or booking time at some of the observatories around the world that allow amateurs to use their facilities. Amateur astronomers can choose a specific area of the night sky to observe regularly, documenting any changes or abnormalities. You never know when you might discover a supernova, new planet or asteroid.

© Matt Gibson / Shutterstock

Dark Places

While many people have never seen a dark night sky that is unaffected by light pollution, doing so has a powerful impact on us both psychologically and physically. Unfortunately, dark skies, like many natural resources, are increasingly at risk due to human development around the world. Over half of the world's population currently lives in urban areas, and the number is expected to grow by another 2.5 billion people in the next few decades. Yet dark places remain, spots where the show in the heavens doesn't have to compete for attention and visitors can relish in the wonder above. Whether you simply feel drawn towards the vast expanse of the shimmering Milky Way above or already love observing the stars and even own your own telescope, seeking out dark skies across the world is a pursuit with many rewards. Chase the Southern Cross down under, watch the revolving stars reflected in Bolivia's salt flats, or study the tapestry above Mt Bromo in Indonesia, these locations will thrill travellers, be they dedicated astrotourists or astrology-obsessed seekers.

Protecting and preserving our dark skies has inspired several movements and associations, including the International Dark-Sky Association (IDA; http://darksky.org). The IDA works with communities and governments to help designate dark-sky places around the world, ensuring they stay dark for future generations. As light pollution threatens to overwhelm our ability to see the night sky at home, it's important to protect areas of darkness we can visit to enjoy the vista above, in the same way that national parks provide access to nature for urban dwellers and suburbanites around the world.

While it may seem as if more light is a good thing for safety and security – especially in cities – light is only advantageous when used in the right way. Light pollution is generally defined as light that affects the darkness beyond the intended area of illumination. For example, turning on car headlights to illuminate the road is not light pollution; shining a light into a back garden when no one is out there is light pollution. Many businesses and homes use excess light or do not efficiently point light where it is needed (and only where it is needed), creating light pollution. The pollution from light affects our ability to see the night sky above us, and urban areas are the worst places for light pollution. For city dwellers, it's common to see fewer than a dozen stars in the night sky. This is a huge impact from light pollution, as thousands of stars can potentially be seen in a dark sky!

Light pollution is more than an annoyance for those who love stargazing. As the amount of light at night has increased, researchers have discovered both physical and psychological effects from what some researchers call the 'loss of night'. Physically, we sleep far worse in places with too much light. Our brains and bodies are built to be diurnal, meaning many of our hormonal and other physical systems operate based on light and darkness. When the sun rises, we naturally wake up; when the sun sets, we grow tired and fall asleep. When we try to sleep with too much light, such as in cities, we may struggle to fall asleep, be unable to reach deeper levels of sleep, be more disrupted during our sleep, and wake up feeling more tired. This affects our lives during the day, decreasing our energy and productivity, and affecting our mood.

The field of ecopsychology looks at the impact of nature on human psychology, including the impact of light pollution and darkness on our well-being. The lack of darkness and poor sleep can actually make us more depressed. We might think that light makes us happier, but at night the opposite is true.

Research suggests that we may feel less of a sense of wonder if we cannot see the night sky, and we may have a decreased perspective of the world and our place in the universe. Some research even suggests that without the ability to go out and see the night sky, we experience a sense of detachment from others and a loss of community. While the psychological impacts of light pollution are less studied and documented than the environmental effects, it's certainly the case that there are impacts from losing the darkness.

Being able to see dark skies is good for us. It helps our bodies, brains, and communities to have access to dark skies. For many city dwellers it may seem like finding dark skies is impossible, but there are some dark-sky locations even near major metropolitan areas, as well as ones tucked into the most remote and beautiful corners of the globe. As more travellers seek out the night sky, there are increasing opportunities for tourism and development for the nearby communities and local businesses.

In this section you'll explore some of the darkest places around the world, from the deserts of the Middle East and Southern Africa to the mountains of Southeast Asia and Oceania. Some destinations are special to amateur astronomers as the sites of annual star parties; in others, tourists may pass through the region by day unaware that they are treading on sacred ground for stargazing. Whether you are seeking out the most pristine dark-sky sanctuaries hidden on remote islands across the globe or merely want to take in a view of the Milky Way on your travels, the night sky is waiting for you to come see the wonders beyond. Visit the dark-sky parks in your own backyard or make a pilgrimage to the unique locations in this section, from the world's deserts to its mountains, all illuminated by the light of the stars above.

Many of the locations in this section have been recognised and designated as dark-sky places by the International Dark-Sky Association. Founded in 1988, the IDA works with locations to help them plan and implement lighting-control policies to preserve the night sky. Four main tiers of IDA designated dark-sky places are in this book:

- Dark Sky Communities, residential areas, cities or towns that implement lighting control;
- Dark Sky Parks, public or private spaces protected for natural conservation that offer good night-sky education and use pollution-reducing lighting;
- Dark Sky Reserves, areas of dark skies surrounded by habitation that reduces light pollution to protect the core zone;
- Dark Sky Sanctuaries, remote locations where skies are naturally dark and need to be protected.

Uluru (Ayers Rock)

AUSTRALIA

One of Australia's most recognisable and revered landmarks, Uluru (Ayers Rock) rises from the Australian outback in the Northern Territory. This giant sandstone formation was pushed up from the Earth's crust over 500

© swissmediavision / Getty Images

Stars above Uluru (Ayers Rock) at night.

million years ago and ever since then has been deeply significant to the Aboriginal peoples of Australia. Called Uluru in the languages of those peoples, it was given the second moniker of Ayers Rock in the late 19th century by an Australian surveyor, and since 2002 the site has gone by the dual names of Uluru (Ayers Rock). Ownership of the site was transferred back to Australia's Aboriginal Anangu people in a symbolic 1985 handback. Its saturated rusty orange hue is bold enough to compete for the attention of even the most dedicated stargazer.

Though it's not easy to visit Uluru (Ayers Rock) from major cities in Australia (Alice Springs, the closest major settlement, is five hours away), it is one of the most popular destinations in the outback's Red Centre, and some 500,000 visitors make the journey each year, drawn by the magnetism of Uluru (Ayers Rock) towering over the outback. In the past, visitors climbed the rock itself, but the practice has been banned to honour the sacred importance of the site. More ambitious visitors might take a flightseeing tour to view it from the air, but all who are able should try circumnavigating the 5.8-mile (9.3km) base.

Visiting Uluru (Ayers Rock) by night to soak in the majesty of this monolith under the breathtakingly clear stars above is a treat. Ayers Rock Resort is the closest accommodation, and offers guided astronomy tours that include a stargazing session. The resort is committed to training Aboriginal employees, and a job is guaranteed for any indigenous Australian who wants one.

The 2028 solar eclipse will pass over Uluru (Ayers Rock), but unfortunately not at totality. Instead, head northeast: the path of totality crosses the Northern Territory along National Hwy 87 north of Willowra to Tennant Creek. You can also visit two impressive impact crater sites, Gosses Bluff and Henbury Meteorites Conservation Reserve, a cluster of 12 small craters formed after a meteor fell to Earth 4700 years ago. The largest crater is 180m wide and 15m deep, surrounded by beautiful country.

Important Info

When to visit: The winter months of May to September are the most optimal for visiting Uluru (Ayers Rock) and the Red Centre.

Website: *https://parksaustralia.gov.au/uluru/*

Warrumbungle National Park

AUSTRALIA

From left: The Anglo-Australian Telescope (AAT) at Siding Spring; Milky Way setting over Siding Spring Observatory.

While there may be national parks in New South Wales that are closer to Canberra, Sydney and Brisbane, none is darker than Warrumbungle National Park. A drive from one of these major cities can take anywhere between 5½ and 8 hours, but Warrumbungle National Park draws visitors who want to experience the natural wonders of the park by day – and by night. Dark-sky preservation efforts and the lack of light pollution earned Warrumbungle its designation in 2016 as the first Dark Sky Park in Australia.

By day, visitors can explore the

© cbphoto / Alamy Stock Photo ; © National Geographic Image Collection / Alamy Stock Photo

Warrumbungle Mountain Range, where hiking, rock climbing and rappelling (abseiling) are popular adventurous activities. There are also several opportunities in the national park to see wildlife such as kangaroos and koalas, and to walk through canyons and caves throughout the park's volcanic formations. Four campgrounds in Warrumbungle are useful for those who want to sleep under the stars. At night it's possible to stargaze in Warrumbungle National Park, especially if you're staying overnight at one of the campgrounds. Camping permits can be obtained at the visitor centre and are required to guarantee you a spot.

Another opportunity to see the stars is at Siding Spring Observatory, located in the western part of the park. The facility is home to several telescopes, including the 13ft (4m) Anglo-Australian Telescope. Siding Spring is operated by the Australian National University Research School of Astronomy and Astrophysics, and is primarily focused on research rather than public outreach or education. Travellers are welcome at the visitor centre by day, and the observatory hosts an annual event, StarFest, each October when the public can visit, meet astronomers and participate in a solar viewing at one of the top astronomical facilities in the country.

In 2028 Warrumbungle National Park will fall inside the path of totality as a solar eclipse passes across Australia. No events have been announced yet, but mark your calendars; the park is likely to sell out camping spots and accommodations in the nearby towns up to a year in advance.

Important Info

When to visit: Spring (September to November) is an ideal time to visit Warrumbungle. September is the driest month, and blooming wildflowers throughout the park add even more colour to your experience. By December both the rainy season and hot summer days make it less pleasant to visit.

Website:
www.nationalparks.nsw.gov.au

Salar de Uyuni

BOLIVIA

High in the mountains near the crest of the Andes, several prehistoric lakes have come together to form the Salar de Uyuni, the world's largest salt flats. The Salar de Uyuni in Bolivia has become an increasingly popular tourist destination, as travellers have become aware of the unusual topography of this picturesque sight via social media. When the surface is dry, the salar is a pure white expanse of the greatest nothing imaginable – just blue sky and white ground. When there's a little water, the surface perfectly reflects the clouds and the blue altiplano sky, and the horizon disappears.

Visiting Salar de Uyuni is a complex undertaking: tour operators offer choices from day trips to a four-day road trip through the 4086-sq-mile (10,583-sq-km) flats, and most travellers typically book an all-inclusive multi-day tour. While it's possible to explore Salar de Uyuni on your own, it's advisable to go with a tour provider due to the size of the flats and the logistics of booking accommodations and amenities.

There are increasing opportunities to visit the Salar de Uyuni at night to take advantage of its dark starry nights and wide horizon. As the Salar de Uyuni is relatively undeveloped (except for intermittent tourism amenities throughout the flats), visitors can experience truly dark skies. Tour operator Ruta Verde runs special itineraries specifically focussed on stargazing, though other providers also serve the area. Stargazing tours to Salar de Uyuni typically depart from the towns of Uyuni or Colchani.

Some operators pair viewing the sunrise or sunset with a stargazing session, and tours typically range from two to six hours in length. Having a tour provider will ensure that you don't get lost in the flats; you'll also have a guide to explain the astronomical sights, which include meteor showers, visible planets, the Milky Way, and the Magellanic Clouds. Given how flat the salt flats are, light travels far, and you'll need to give your eyes time to adjust once you arrive at your stargazing spot.

Clockwise from top: Under the sky in Bolivia; Salar de Uyuni reflecting the clouds; Valle de la Luna in Bolivia.

Consider adding the Valle de la Luna (Valley of the Moon) to your itinerary in Bolivia. The unusual geological formations in this area are outside the Bolivian capital of La Paz and are an easy stop if you're visiting northern Bolivia to see the city or Lake Titicaca. Isla Taquile, an island reachable by boat from Puno in Peru across Lake Titicaca, has wonderful dark skies too and offers homestays.

Important Info

When to visit: The most popular time to visit is during the rainy season, December to March, when water pools on the salt flats and turns the ground to a mirror of the sky. This is your best chance to shoot a photo with stars overhead and reflected in the water below.

Website: *www.rutaverdebolivia.com/tour/uyuni-stargazing-tour*

Parque Nacional dos Lençóis Maranhenses

BRAZIL

Formed by sedimentary river deposits and equatorial winds from the Atlantic Ocean, Parque Nacional dos Lençóis Maranhenses is a sea of sand on the northeast coast of Brazil and one of the country's most unusual national parks. The 598 sq miles (1549 sq km) of sand dunes may look like a desert, but the area receives almost five times more rain than the average desert.

It's this 47in (1.2m) of rain per year that turn the sweeping dunes of Parque Nacional dos Lençóis Maranhenses into an eco-adventure destination. Each year rain settles between the dunes, creating freshwater lagoons that persist for months and support a unique range of plants and animals, including coastal, desert and savannah species. Seemingly a desert at first glance, the park's distinctive combination of biological and ecological conditions make it a uniquely diverse location.

Because of the annual dance of wind, sand and water, Parque Nacional dos Lençóis Maranhenses is relatively undeveloped beyond two oases in the dunes. This has helped preserve the night sky above the park, so it's easy to go stargazing while adventuring here. The park's location just 2.5° below the equator also affords you the opportunity to see a night sky view that includes both northern and southern constellations. In addition to hiking, ATV and horseback tours, and canoeing or swimming in the temporary freshwater pools, surfing is another option: the park is located along 43 miles (69km) of Brazilian coastline. It's a wonderful detour from a trip into the Amazon Basin, which abuts the park.

Clockwise from left: Lagoon and dunes at dusk; star trail above the dunes; a freshwater lagoon.

The first modern-day observatory in the Americas was built on Antonio Vaz island in Brazil by the Dutch in the 1630s. Astronomer Georg Markgraf observed the solar eclipse from this observatory in 1640, but unfortunately the building was destroyed in 1654. Today you can visit the Malakoff Tower in Recife, which houses a modern observatory, or the Space Science Museum in nearby Olinda to learn about astronomy in Brazil.

Important Info

When to visit: The rainy season (January to May) produces the freshwater lagoons in the dunes. The lagoons dry up between June and November, so visit in June, July or August.

Website: *http://visitbrasil.com/en/atracoes/lencois-maranhenses-national-park.html*

Jasper National Park

CANADA

Independent of the International Dark-Sky Association, some countries choose to enact their own standards of dark-sky preservation. Canada is one such country, and Jasper National Park is one of 10 dark-sky preserves throughout Canada (to learn more about the biggest dark-sky preserve in Canada, Wood Buffalo National Park, see p172). Known by day for its picturesque Canadian Rocky Mountains, colourful alpine lakes and stunning sunsets, Jasper National Park beckons all travellers

© Stocktrek Images, Inc. / Alamy Stock Photo; © Krishna.Wu / Shutterstock

Above: The Columbia Icefields and Athabasca Glacier in Jasper. Opposite: Spirit Island at Maligne Lake.

who love the outdoors. It also welcomes avid astrotourists who seek out dark skies after a day of adventure.

A four-hour drive from Edmonton, Alberta, or a five-hour drive from Calgary, Jasper National Park is a large and relatively accessible protected area for all kinds of outdoor activities. Hiking, cycling, mountaineering and camping are common summer activities, and when the weather turns cold, visitors enjoy skiing, snowshoeing and even ice climbing. Increasingly, visitors flock to Jasper National Park during warm-weather months for day hikes and photography. That it offers something for almost every kind of traveller is one of the main reasons Jasper National Park ranks consistently among the most-visited destinations in Canada.

By night the sky above Jasper's iconic mountains and lakes steals the show. Limited light pollution – ensured by the park's Dark Sky Preserve status – guarantees a clear view of the stars, Milky Way and even the aurora during dark winter months. Each October the park holds the Jasper Night Sky Festival (*https://jasperdarksky.travel*), one of the largest of its kind in the world. The festival brings speakers, astronomers, night sky advocates and the general public together to learn about and experience the night sky. It also includes food events, a symphony under the stars and astrophotography sessions.

The Night Sky Festival in Jasper is one of many similar celebrations across Canada. Throughout the summer months and into September, national parks host dark-sky festivals. These include Kluane's Annual Dark Sky Festival in the Yukon Territory, Milky Way Day at Elk Island National Park in Alberta, dark-sky nights at Point Pelee National Park in Ontario, and the Night Sky Celebration at Terra Nova National Park in Newfoundland and Labrador.

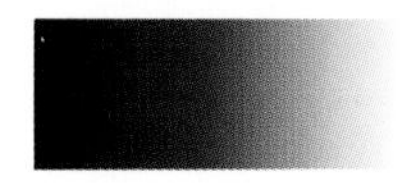

Important Info

When to visit: The best times to visit for dark skies are the winter months of November through February. If snow sports aren't your thing, the shoulder seasons of September to October and March to May also have dark nights but with less snow.

Website: *www.pc.gc.ca/en/pn-np/ab/jasper*

Mont-Mégantic

CANADA

Mont-Mégantic paved the way and became a model for dark-sky protection as the world's first designated Dark Sky Reserve. Located in southern Québec near the US border with New Hampshire and Vermont, Mont-Mégantic Dark Sky Reserve encompasses some 3300 sq miles (8547 sq km), including the community of Sherbrooke, Mont-Mégantic National Park, and the observation facilities therein.

Mont-Mégantic National Park is a popular hiking destination but also great for snowshoeing and cross-country skiing in winter months. Paragliders and hang gliders also love to catch a ride off Mont St-Joseph, whereas mountain bikers and cyclists appreciate the challenge of navigating the mountains throughout the park, and birders enjoy spotting some of the avian species protected here. Camping and other rustic accommodations are available in the park, where you can set up on your own for some stargazing.

In addition to experiencing the national park by day, your main must-see stops are the Mont-Mégantic Observatory, the Popular Observatory, and a visitor centre, ASTROlab. Mont-Mégantic Observatory operates the second-largest telescope in eastern Canada, 63in (1.6m) in diameter. The observatory is open to the public for daytime tours and night-time astronomy events.

ASTROlab's museum and activity centre is the focal point for space enthusiasts visiting Mont-Mégantic. Indoor exhibits and outdoor viewing instruments provide a perfect opportunity to learn more about the stars. Astronomy Night events combine an astronomy lecture with guided observation through telescopes and giant binoculars. Other regular events include evening viewings at the Popular Observatory, plus there are special events such as the annual Astronomy Festival in July and the Perseids Festival in August, which focuses on viewing the year's most active meteor shower.

From top: The park is a snowy wonderland for much of the year; Parc National du Mont-Meganic and its Observatory.

The observatory at Mont-Mégantic is along the centre line path of totality for the 2024 solar eclipse across North America. While no events have been announced yet, plan at least six months ahead to book accommodation near Mont-Mégantic National Park so you can see the stars pop out during totality. Montreal will be able to see totality as well, but for a shorter duration, as the city is on the northern edge of totality's path. And eclipse or not, this is a great destination for bird watching.

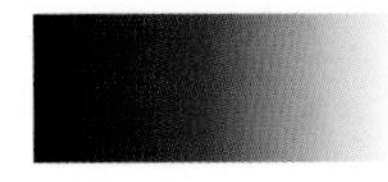

Important Info

When to visit: The Perseids Festival occurs mid-August each year to view the spectacular meteor shower.

Website: *http://astrolab-parc-national-mont-megantic.org*

Parque Nacional Torres del Paine

CHILE

Soaring almost vertically above the Patagonian steppe in Chile, the granite pillars of Torres del Paine (Towers of Paine) dominate the landscape of South America's finest national park. Part of Unesco's Biosphere Reserve system since 1978, this 698-sq-mile (1808-sq-km) park is, however, much more than its one greatest hit. Its diverse landscapes range from teal and azure lakes to emerald forests, roaring rivers and the radiant blue glaciers of the Southern Patagonia Ice Fields, including Tyndall and Grey glaciers. Guanacos roam the vast open steppe, while Andean condors soar beside looming peaks.

The rugged mountains and gleaming glaciers draw adventurers from around the world to hike, climb and kayak. Several famous hiking routes traverse Torres del Paine, among them the five-day 'W' route and the eight-day circle route around the Paine Grande and Cuernos del Paine. Although this is a mostly undeveloped area, in part due to its protected status, there are a handful of vista points, camping areas and lodges to service the 250,000 visitors a year. One main road with several spurs provides entry to the park.

Based on its geography and altitude, Torres del Paine has a rainy autumn (March and May) and a cold, potentially snowy winter (June to September). Most travellers visit during the spring and summer months which coincide with longer days – including several weeks when it gets no darker than twilight between late November and mid-January. To take advantage of the limited darkness available in months when the sky is clear requires advance planning and the willingness to stay up into the wee hours. It's worth it, though: Torres del Paine provides a stunning natural foreground against which you can view the Milky Way, the Magellanic Clouds, and the southern sky.

Clockwise from top: Dawn moon at Torres del Paine; ice trekkers on Grey Glacier; a couple at the famous three towers.

Torres del Paine is in the path of the December 2020 solar eclipse but will only experience a 60% partial eclipse. Plan ahead and bring eclipse glasses if you're visiting during this time and want to view the partial eclipse. Additionally, the southern lights (aurora australis) have been spotted in Patagonia from time to time. To have the best chances of seeing this rare sight, plan to visit during the darkest months of the year – but be prepared to face wind, rain and snow while trying to glimpse them.

Important Info

When to visit: The shoulder months (October and April) are the best times to visit for stargazing, since you'll get to experience dark nights without wintry weather.

Website: *www.parquetorresdelpaine.cl/es*

Eifel National Park

GERMANY

While Eifel National Park may be young by national park standards – it was only established in 2004 – it has an ambitious long-term goal: to return major portions of this German parkland to wilderness. To this end, the Eifel National Park Authority has established a comprehensive plan for visitor management that allows travellers from around the world to experience the forest while also protecting the natural resources that will help return the countryside to its primitive state, dark skies included.

Eifel National Park covers 42 sq miles (109 sq km) in west-central Germany along the Belgian border. It is estimated that nearly 20 million people live within two hours of the park. While this is great for encouraging European residents to explore the wilderness by day, it poses a challenge for preserving darkness. Eifel National Park was designated as a provisional Dark Sky Park in 2014, and work continues to reduce light pollution and ensure low-light emission compliance where possible.

By day Eifel National Park is a natural playground. Over 150 miles (242km) of paths are available for hiking, cycling and even cross-country skiing in winter. The park is accessible to travellers with a wide range of abilities, and park rangers offer regular guided walks and talks to introduce visitors to the park.

After sunset, park rangers provide additional information on the wonders of the night sky. The Vogelsang observatory, located on property reclaimed from a former Nazi estate, educates visitors from around the world. The observatory also has an astronomy workshop, Stars Without Borders, that introduces the basics of astronomy and points out notable objects in the night sky as well as the Milky Way (when visible). A variety of tour operators are beginning to offer guided tours in Eifel National Park too, increasing the options for travellers who want to see the night sky during their visit.

Clockwise from top: Scenic Bertradaburg nearby; Gemündener Maar volcanic lake in the Eifel; hiking Eifel.

© art4stock / Shutterstock; © Horst-Koenemund / Shutterstock; © Panther Media GmbH / Alamy Stock Photo

Harald Bardenhagen both helped establish Eifel National Park as a Dark Sky Park and helped design the Stars Without Borders observatory programming for the Vogelsang observatory. 'To see a starry night sky – a very fundamental, direct nature experience – this really touches the heart and the brains of the visitors. I want visitors to rediscover the value of the darkness, and I want them to become ambassadors for a natural, starry night sky', says Bardenhagen.

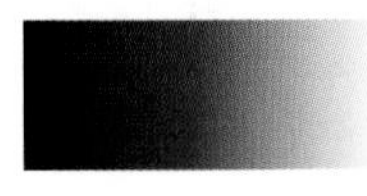

Important Info

When to visit: Eifel is good all year, but in the four weeks surrounding the June solstice, the sky doesn't get fully dark, which may impact stargazing opportunities.

Website: *www.nationalpark-eifel.de/en*

Westhavelland Nature Park

GERMANY

You may think it takes hours of travel to reach dark skies around the world, but as proven by Westhavelland Nature Park in Germany, it's possible to preserve nature and protect dark skies even in relatively close proximity to major cities.

Located 44 miles (71km) east of Berlin, Westhavelland Nature Park was established in 1998 to protect the sweeping marshlands of the River Havel and nearby Gülper Lake and the species of animals and birds that call the wetlands home. At 507 sq miles (1313 sq km), Westhavelland is the largest protected wetland in Europe, and it's home to a wide variety of birds, including several endangered species. Birdwatching and cycling tours are common ways to experience Westhavelland during daylight hours.

Once the sun sets, it's immediately obvious why Westhavelland was named Germany's first Dark Sky Reserve: the relative lack of development in the flood-prone area has protected the night sky and makes this one of the best places for those living in or travelling to northern Germany to see the Milky Way. It's estimated that nearly six million people in the Berlin-Brandenburg region can easily access Westhavelland Nature Park and enjoy the dark skies here.

Each September the park hosts the family-friendly Westhavelland AstroTreff, a multi-day star party where you can learn astronomy, stargaze, camp beneath the stars and participate in solar observation during the day. It helps if you speak German when attending this or less formal star parties at the nearby Sternenblick Parey observatory, as none of the events are conducted in English.

You can get to Westhavelland from Berlin by car in less than two hours; few bus or train options reach this part of the country. Rathenow is the largest community in Westhavelland Nature Park; Brandenburg sits along the southern border and has all major travel amenities.

From top: Westhavelland sky at dusk; Berlin's skyline with Alexanderplatz TV tower, a mere two hours from the nature park.

You can see more than just the stars here. Thomas Becker works as an official in the nature park and remembers viewing galaxies through a telescope one night in April. 'Quite unexpectedly, the aurora appeared', Becker recalls. 'Thanks to the dark sky in the Nature Park, it was very clear. Somewhere else it would probably have been lost in the haze.' While seeing the aurora isn't common so far south in Europe, at Westhavelland Nature Park it's possible under the right conditions.

Important Info

When to visit: May and September are the ideal mix of warmth and the highest chance of clear skies. Summer gets humid, and winter can be freezing.

Website:
www.sternenpark-westhavelland.de

Hortobágy National Park

HUNGARY

From left: Common cranes (Grus grus) in flight at sunset; the park is also home to Przewalski's horses.

As dark-sky preservation and certification increase around the world, countries are beginning to take pride in the number of their dark-sky sites. One such destination is Hungary, home to three certified dark-sky places, including Hortobágy National Park to the east.

Sweeping grasslands and marshlands, like those in Hortobágy National Park, are ideal areas for stargazing. The flat land without mountains or other geological formations and the relative absence of tall plant species ensure that you have an almost unobstructed view of the night sky here. As the first national park in Hungary, Hortobágy is a relatively undeveloped region of the European continent, protecting

© Ferenc Cegledi / Alamy Stock Photo; © Peter Horvath / Getty Images

pristine night skies and nocturnal species. For this reason, part of Hortobágy was certified as a Dark Sky Park in 2011, and efforts have been made to retrofit lighting in nearby communities to keep the skies dark moving forward.

Hortobágy National Park has a significant value during the daytime as well, as a protected biosphere reserve and Unesco World Heritage Site. The park is home to wild animals such as wolves, wild horses and jackals, and is especially attractive to migratory birds. Travellers visiting Hortobágy can observe species in an ecosystem that has remained mostly undisturbed for thousands of years. Other activities in the national park include learning about the unusual geology that formed the alkaline plains of Hortobágy, and visiting the town of Hortobágy to experience the heritage of herdsmen and their grazing animals in the region. It's also the best spot to try Hungarian crepes à la Hortobágy, the local specialty pancakes filled with minced meat and covered in a paprika sauce.

At night Hortobágy offers a variety of astronomical activities. The Fecskeház Youth Hostel and Field Study Centre has an observatory dome and offers guided viewings on clear, dark nights; the hostel has space for 34 to spend the night. National park employees also organise stargazing walks and astronomy talks to educate visitors on the wonders overhead.

István Gyarmathy, dark-sky coordinator for Hortobágy, understands the importance of the night sky to all living creatures in the park: 'You can see the natural change of light during twilight and then experience the natural sounds and the heavens full with stars at night-time', he shares. 'This inspired the shepherd culture since ancient times too, and the relationship to the starry sky is still part of the living shepherd tradition.'

Important Info

When to visit: The best times to visit are the shoulder seasons of spring (April and May) and autumn (September and October), when warm days turn into not-too-cool nights perfect for stargazing and spotting meteor showers.

Website: *www.hnp.hu*

Zselic Starry Sky Park

HUNGARY

From left: Aerial view of the astral observatory at Zselici Csillagpark; Ban Jelačić square in Zagreb, Croatia.

As is the case for most major European cities, it is difficult to find dark skies directly near Budapest, Hungary. Yet two hours southwest, towards the border with Croatia, Zselic Starry Sky Park, or Zselici Csillagpark in Hungarian, is one of the best places for stargazing in eastern Europe. Designated in 2009, it was one of the first Dark Sky Parks in Europe. Zselic Starry Sky Park is located within the Zselic National Landscape Protection Area and the staff help ensure that the mission to protect natural resources is met by day and night.

Though off the beaten path for most European travellers, the park has a lot to offer to anyone who wants to enjoy the pristine night skies. In addition to an observatory with viewing telescopes, there's a planetarium with regular programming for all ages, an exhibition on astronomy and nature, and a meteorite collection. Visitors can

© Attila Fodemesi / Shutterstock; © Shevchenko Andrey / Shutterstock

climb a five-storey viewing platform to get even closer to the stars. The facility, exhibits and planetarium are open for daytime visits Tuesday through Sunday; the park also offers solar observation sessions during the day. Night-time observatory programmes on Fridays focus on different objects visible in the night sky throughout the year.

You can also learn about nature and the ecosystem in this part of the world during a daytime visit. As part of the Zselic National Landscape Protection Area, the staff at Zselic Starry Sky Park offer regularly scheduled guided walks and lectures about flora and fauna in the area, and they relate these educational experiences to nocturnal animal behaviours and the importance of protecting the night sky. The local Hotel Kardosfa has installed fully compliant fixtures to reduce light pollution and works as a great base for an overnight or multi-day stay in the area. If visiting Croatia before or after your time in Hungary, consider adding a stop at Zagreb's observatory in the city's Old Town, known as Popov Toranj (Priest's Tower) and founded in 1903. The Zagreb Technical Museum also features a planetarium. Closer to the Croatian coast, the Višnjan Observatory is responsible for the discovery of hundreds of exoplanets and asteroids.

In Hungarian mythology from the early Magyars, the night sky is a big tent held up by the Tree of Life which connects the upper and middle worlds. In this tent, you can see the sun, moon and other celestial objects in the upper world, also inhabited by the gods and good souls. The stars were thought to be holes in the tent, letting light through. At the top sat the messenger hawk, Turul. Archaeoastronomers are now reconstructing these lost myths.

Important Info

When to visit: Zselic Starry Sky Park offers the chance to see the phenomenon of zodiacal light, a faint elongated light is thought to be the reflection of sunlight from particles of ice and dust within the plane of the solar system, most often visible in spring and autumn.

Website:
http://zselicicsillagpark.hu

Jökulsárlón

ICELAND

In many ways Iceland is a dream destination for travellers who want to feel like they've left this planet. It has been used as a filming location for other worlds in movies such as *Interstellar* (2014), thanks to its unusual geological formations left by millennia of volcanic and glacial impact on the island. Throughout the country you can find places that leave you speechless and in awe of our planet – or wondering if you've left it entirely for some other realm.

One such place is Jökulsárlón glacial lake (pronounced yokul-sar-lon), located in southeastern Iceland between the edge of the Vatnajökull ice cap and the Atlantic coast. Formed by the glacier Breiðamerkurjökull as it recedes back into Vatnajökull, this massive lake is easily accessed while driving the famous Ring Road route around Iceland. Although it looks as though it's been here since the last

© Franckreporter / Getty Images; © Myron Standret / Alamy Stock Photo

ice age, the lagoon is only about 80 years old. Travellers often stop to take pictures with the huge icebergs that fill the lake year-round, or use accommodations in the area as a base while they book a guided tour to Vatnajökull itself. By day, visitors can explore the stunning ice caves in Vatnajökull, which offer great photo opportunities and give a sense of the massive power of glaciers to shape Iceland over time. There are also boating tours on Jökulsárlón that you can enjoy in daytime between your stargazing sessions.

Jökulsárlón is a popular destination for hunting the northern lights along Iceland's southern coast too. Even when the northern lights aren't visible, it's a good spot to go stargazing.

From left: The aurora borealis above Jokulsarlon glacial lake; the icy lagoon under the Milky Way.

It's common to see the Milky Way, meteors and planets here. As at other dark-sky locations near water, when the conditions are right you may even spot the reflection of stars in between icebergs in the glacial lagoon. There are no established stargazing tours to Jökulsárlón, but you can easily set out on your own when visiting the area given the convenient location off the Ring Road. Observe the night sky above glacial icebergs on their journey towards the ocean, as they refreeze and sometimes topple with a crash on a trip that can last five years.

The 2026 total solar eclipse will pass over Iceland. Totality will be visible from Reykjavík and parts of western Iceland, while Jökulsárlón will experience a 97% partial eclipse. This may seem like a total eclipse, but it's not quite totality, so be sure to wear eye protection and practice safe eclipse-viewing techniques. If you choose to view the partial eclipse at Jökulsárlón, be prepared for some great photography opportunities in the expansive, 250m-deep lagoon.

Important Info

When to visit: To go stargazing or to view the aurora at Jökulsárlón, you'll need to visit during the darker months of the year or stay up late.

Website:
http://icelagoon.is

Mt Bromo

INDONESIA

Located in the fiery heart of East Java, Mt Bromo is considered one of the top attractions in Indonesia. Travellers come to climb near this active volcano to watch the sunrise, a spectacular display of colours in the morning sky, as well as to explore the surrounding caldera and other active volcanoes. Mt Bromo is part of the Bromo-Tengger-Semeru National Park, which encompasses some 300 sq miles (777 sq km), including five volcanoes and the Tengger 'sand sea'. By day, travellers can hike in the national park or book a guided 4WD tour to see these unusual geological features.

Mt Bromo is quietly gaining renown as a great stargazing location in Indonesia, as astronomers observe and astrophotographers capture photos of the Milky Way, Andromeda Galaxy and the Magellanic Clouds. Located less than 10° south of the equator, Mt Bromo and the surrounding region are also a good destination for viewing the southern night sky and meteor showers like the Southern Taurids.

Unlike other dark-sky places that offer designated locations and planned astronomy activities, stargazing in Mt Bromo is a mostly independent affair. Travellers should be flexible to accommodate the weather and prepared for cold temperatures once you start hiking up the mountain. Additionally, parts of the national park and Mt Bromo are occasionally closed due to volcanic activity, a hazard of its location on the Pacific Ring of Fire, so be sure to check with local park officials before setting out at night. Consider planning an early morning stargazing session so you can also view the sunrise from Mt Penanjakan, a popular lookout with Mt Bromo in the foreground. Visitors can also plan their trip around the annual Yadnya Kasada ceremony, an offering from the Tenggerese people to the mountain in honour of a Majapahit kingdom legend.

Clockwise from top: A meteor shower above Bromo Tengger Semeru National park; Mt Bromo under the Milky Way; exploring in Mt Bromo by 4WD.

© Witthaya Prasongsin / Alamy Stock Photo; © Nuttawut Uttamaharad / Alamy Stock Photo; © Aws Zuhair / Alamy Stock Photo

If your travels take you to West Java in addition to East Java, be sure to plan a trip to Bosscha Observatory on the outskirts of Bandung. Bosscha Observatory is the primary observatory in Indonesia and is home to five telescopes including a 24in (61cm) Zeiss double refractor. You can book a daytime or night visit to the Bosscha Observatory on their website. Bandung is a 3½-hour drive from Jakarta, where most international flights arrive and depart.

Important Info

When to visit: Summer is the high season, and has the driest months of the year, ensuring clear, dark skies. By September, night-time temperatures begin to drop.

Website: *www.indonesia.travel/gb/en/destinations/java/bromo-tengger-semeru-national-park/mount-bromo*

Kerry Dark Sky Reserve

IRELAND

From left: Uragh Stone Circle, Gleninchaquin; the rocky coastline on Coumeenoole Bay near Slea Head.

Between the rocky North Atlantic shores and the mountains of Kerry, the Ring of Kerry on Ireland's west coast cuts a path through some of the darkest skies in the northern hemisphere. Here you'll find Kerry Dark Sky Reserve, designated as a reserve by the International Dark-Sky Association in 2014 and considered one of the best places for travellers seeking dark skies in Ireland.

Travellers visiting the Ring of Kerry or the longer Wild Atlantic Way for the striking castles and intriguing stone circles may not even be aware that they're right next to a premier dark-sky

© David Epperson / Getty Images; © Pete Seaward

site; the open spaces and rural Irish countryside are beautiful by day and help preserve dark skies once the sun goes down. Kerry Dark Sky Reserve is roughly 270 sq miles (699 sq km) and encompasses several small communities. For the most part, though, there's little light pollution in the core zone – even from the town of Killarney, which is only 40 miles (64km) away. At night even trained astronomers struggle to differentiate the constellations amid all of the stars in the sky here.

Kerry Dark Sky Reserve is one of the best examples of a stargazing location that recognises the allure of the dark sky as a draw for travellers. In addition to daytime tours that highlight the wildlife, geography and history of the county, the Kerry Dark Sky Reserve offers trained stargazing guides who help you take advantage of the sights after dark. These excellent guides highlight the constellations, the science of modern astronomy and the importance of astronomy in Neolithic culture, which played a role in many of the stone circles and ruins throughout magical County Kerry.

The Kerry Dark Sky Reserve tourism office in Dungeagan offers resources and information on visiting the more than 20 parking areas throughout the reserve where you can stargaze easily on your visit.

If you're travelling northward on the Wild Atlantic Way from Kerry Dark Sky Reserve, be sure to plan a stop at Ballycroy National Park and the Wild Nephin Wilderness along the Bay Coast. In 2016 these two protected areas were jointly designated as Mayo Dark Sky Park. By day you can learn more about this unique bog and peatland, and by night enjoy viewing the sky from the Claggan Mountain Boardwalk or Letterkeen Bothy. On clear nights 4500 stars are visible.

Important Info

When to visit: Rain and cloud cover are common in this part of Ireland. The best months for clear skies are July through September.

Website: *http://kerrydarksky.com*

Makhtesh Ramon

ISRAEL

While Makhtesh Ramon may be called the Ramon Crater in English, this fascinating geological feature in Israel's Negev desert is actually not a crater at all. It was not created by a meteor impact or a volcanic eruption; instead it was formed over millions of years through erosion. Once covered by an ocean, later river movements through the Arava Rift valley carved out the crater that exists today, remarkable for its vibrant clay hills and surrounding mountains. Though not of celestial origin, the sweeping landscape of Makhtesh Ramon is in a lesser-developed part of southern Israel, and as such has great stargazing opportunities borne of low light pollution. In 2017 Makhtesh Ramon was designated as a Dark Sky Park, the first of its kind in the Middle East.

As the largest erosion crater in the world and Israel's largest national park, Makhtesh Ramon is also a wonderful place to view the unique flora and fauna of the region, from foxes, gazelles and leopards to wild horses. A visitor centre in the town of Mitzpe Ramon is the starting point to set out and explore the makhtesh, including the small Bio-Ramon facility where you can see some of the wild species close up. Perched on the makhtesh rim, this visitor centre has extremely helpful staff who are willing and able to answer questions about the Makhtesh Ramon Nature Reserve, its habitats and its outdoor activities. Most of the museum serves as a memorial for Israeli astronaut Ilan Ramon, who died in the Space Shuttle Columbia disaster, for those with an interest in space exploration as well as astronomy. The last section and several films focus on the makhtesh's geography and natural history. There are also hiking and cycling trails, rock climbing and rappelling (abseiling) opportunities, and 4WD tours through the desert can be done independently or with a guide. For a unique experience, at Be'erot Campground in the heart of Makhtesh Ramon, you can even spend the night, learn about Bedouin culture and sleep under the stars. A stay includes a meal of traditional food, including fresh pita and sweet tea.

With its desert climate Makhtesh

Makhtesh Ramon is home to a staggering number of species.

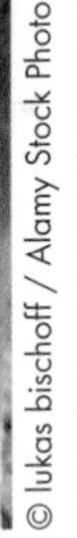

© lukas bischoff / Alamy Stock Photo

Ramon has been left largely undeveloped, which preserves the night sky and reduces light pollution throughout the area. Whether you're travelling independently or with a guide, it's easy to find a place where you can see the stars, Milky Way and visible planets. A select number of tour operators based in Mitzpe Ramon also offer evening stargazing tours if you want a guided look at the night sky. Visitors gazing up at the stars here will experience the thrill of knowing that it's close to the same view Bedouins have been taking in for centuries from this very spot.

Israel Nature and Parks Authority Chief Ecologist Noam Leader says, 'For most visitors a clear view of crisp, unpolluted night skies exhibiting a grand view of the Milky Way and thousands of stars is most likely a rare and awesome view, which unfortunately they cannot enjoy at home due to light pollution. The knowledge that such a view is becoming a rarity presents a great responsibility to conserve it, and is also an opportunity to educate visitors to go back home and ask themselves how they "lost the stars" in their living environment, and try to change this reality.'

Important Info

When to visit: Hot days and cool evenings in summer are perfect for overnight camping and stargazing, but be sure to stay hydrated.

Website: *www.parks.org.il*

Iriomote-Ishigaki National Park

JAPAN

In Japan's southernmost Okinawa Prefecture deep in the East China Sea, the Yaeyama Islands of Iriomote, Ishigaki, Kohama, Kuro, and Taketomi form Iriomote-Ishigaki National Park. These islands are accessible only by plane and ferry. They're closer to Taiwan than to Japan, making the national park a unique aspect of Japan's geography and home to the only subtropical rainforest in the country. In 2018 Iriomote-Ishigaki National Park was also designated as the first Dark Sky Park in Japan due to its remoteness and relative lack of development. Although only 20km west of Ishigaki-jima, Iriomote-jima could easily qualify as Japan's last frontier. Dense jungles and mangrove swamps blanket more than 90% of the island, and it's fringed by some of the most beautiful coral reefs in Japan, making Iriomote-jima a top destination in Japan for nature lovers.

Travellers to Iriomote-Ishigaki National Park can relish the varied terrain and unique species of flora and fauna, and avid snorkellers and scuba divers will be especially pleased by the fish in the tropical waters. Some species protected by the park, including yaeyama palm trees and the Iriomote wild cat (itself nocturnal), are endemic. There are also opportunities for hiking Mt Omoto on Ishigaki or exploring some of the island's picturesque waterfalls, canoeing through mangrove forests, and snorkelling or boating among the coral reefs that surround some of the islands.

While there is human settlement and development on the Yaeyama Islands, only 55,000 people live on the 228 sq miles (591 sq km) of land, primarily on Ishigaki. This means that there are opportunities for cultural experiences by day – and dark night skies on some of the other islands that are far less developed. As part of the effort to receive Dark Sky Park designation, a plan to retrofit and reduce lighting has already reduced light pollution and made Iriomote-Ishigaki National Park among the darkest places in Japan.

© Janelle Orth / Alamy Stock Photo

If you're travelling to Iriomote-Ishigaki National Park to enjoy the night sky, consider adding on a stop to visit one or both of Japan's public launch facilities. Located in southern Japan, Tanegashima Space Center and Uchinoura Space Center are open to the public on non-launch days, where you can learn more about the Japanese Aerospace Exploration Agency (JAXA) and Japanese space programmes.

Important Info

When to visit: Iriomote-Ishigaki National Park and the Yaeyama Islands are temperate year-round. Rain and cloudy weather are more likely in the winter, while late summer through autumn has clearer skies.

Website: *www.env.go.jp/en/nature/nps/park/iriomote*

The Milky Way and Uganzaki Lighthouse on a clear night.

Wadi Rum

JORDAN

Deep in the Arabian Desert, Jordan's Wadi Rum Protected Area is a landscape that looks like it was transported straight from Mars. Rust-coloured sand dunes spread through valleys formed by massive red rocks, carved over time by the wind to resemble sleeping giants. Once you've seen the moonrise over this beautifully stark landscape, it comes as no surprise that Wadi Rum bears the nickname 'Valley of the Moon'.

Wadi Rum and the surrounding region are not particularly hospitable for permanent settlement, which is why the Jordanian Bedouin lived a nomadic lifestyle for centuries and the area is relatively undeveloped – and unspoiled by light pollution. This makes it a prime destination for viewing the winter night sky, including the constellation Orion, the Hunter, and the brightest star in our sky, Sirius.

In Wadi Rum, a common stop on most itineraries through Jordan, you can enjoy a traditional meal, sleep in a replicated Bedouin tent and ride a camel or 4WD through the desert. Hot days transform into cool nights, and going a few steps beyond the lighted pathways in each tourist camp gives you a great opportunity to stargaze under the inky black sky.

The most extraterrestrial-focused accommodation in the area is Sun City Camp, which partnered with a European designer to create geodesic 'Martian' domes. Inspired by the surrounding landscape and the 2015 film of the same name, which was filmed in the area, these domes offer guests a chance to watch the stars while drifting off to sleep.

Wadi Rum is roughly 60 miles (97km) south of the archaeological site of Petra, Jordan's famous 'Rose City', so named for the hue of its stone. While stargazing from Petra is not as easily done as in Wadi Rum, the famous Petra-by-night tour is ideal for any traveller who enjoys spending time under the stars and learning local folklore and history. Established tour itineraries through Jordan will typically include one or two nights at each of these locations.

Clockwise from top: The Milky Way above Wadi Rum; candles burn at Al-Khazneh in Petra; Wadi Rum by day.

Wadi Rum is Jordan's most popular filming location. It was first popularised and became a tourist destination after being featured in *Lawrence of Arabia* (1962). More recently it has become a favoured location for science-fiction movies, including *Red Planet* (2000), *The Martian* (2015) and *Rogue One: A Star Wars Story* (2016).

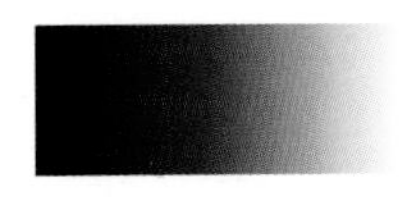

Important Info

When to visit: Jordan has the best weather in November and February to March. Despite hot days, nights in Wadi Rum may be cool, so pack layers if you plan to spend time stargazing.

Website: *http://international.visitjordan.com/Wheretogo/Wadirum.aspx*

Erg Chebbi

MOROCCO

If visiting North Africa is on your list and you're looking for a dark-sky hotspot, Morocco may be a perfect destination. Travellers experiencing the country's eclectic and immersive blend of Middle Eastern, European and African influences come away with intense memories and colourful souvenirs from the medinas. It's also possible to go stargazing on an itinerary through Morocco, making a trip from the major coastal cities to the arid pre-Saharan landscape. Erg Chebbi and its populated centre at Merzouga are a great place to visit for camel riding and learning about traditional life in the desert, as well as for enjoying sweeping views of the stars. Morocco has a rich history of medieval Islamic astronomy, with a notable golden age of discovery in its past. It's also home to the Morocco Oukaïmeden Sky Survey (MOSS) in the country's High Atlas mountains.

From left: Caravan on camels going through the sand dunes in the Sahara desert in Morocco; the Milky Way over the desert.

© Yongyut Kumsri / Shutterstock; © dubassy / Alamy Stock Photo

When a wealthy family refused hospitality to a poor woman and her son, God was offended, and buried them under the mounds of sand called Erg Chebbi. So goes the legend of the dunes rising majestically above the twin villages of Merzouga and Hassi Labied, which for many travellers fulfill Morocco's promise as a dream desert destination. But Erg Chebbi's beauty coupled with Merzouga's accessibility has its price in popularity, with surging visitorship.

The town of Merzouga is located far inland, near the Algerian border. The primary feature of the region is Erg Chebbi, a sea of sand dunes that rise from the rocky desert floor. Most visitors find themselves in Merzouga as part of a multi-day itinerary through Morocco; independent travellers can make the nine-hour drive from Marrakech. In Merzouga there are a variety of accommodation options, from luxury hotels to desert camps. Once in Merzouga, it's common to book a guided tour out onto Erg Chebbi for sandboarding, desert buggy or quad racing on the sand, and camel rides.

Various tour operators offer stargazing-focused excursions, affording the chance to see the stars and stay in a desert camp. Some of these tours are guided, and telescopes may be provided. The skies above Erg Chebbi are still incredibly dark, and the desert here offers one of the best chances to see the Milky Way with an unobstructed view in all directions. Add a stay at the stargazing hotel SaharaSky near Zagora, with its own private observatory.

Important Info

When to visit: Morocco is very hot in summer, and desert nights get cold during winter. Visit in spring (March to May), when desert oases are in bloom, or in autumn (September to October).

Website: *http://moss-observatory.org/*

NamibRand Nature Reserve

NAMIBIA

Deserts and arid regions are among the best places worldwide to stargaze. Life is harsh here, and this reduces human development and light pollution. In the NamibRand Nature Reserve, protection adds an extra clarity to one of the internationally recognised darkest skies in southern Africa.

NamibRand Nature Reserve, a private conservation initiative, was established in 1984 by JA Brückner, a successful businessman who began acquiring farmland in the Namib-Naukluft National Park. Working with other farm owners, he successfully petitioned that the land be turned into a nature reserve.

This page: Zebra in the desert landscape of the NamibRand Nature Reserve in Namibia.
Opposite: The Namib Desert at night.

© Felix Lipov / Alamy Stock Photo; © Jacob Le Roux / Alamy Stock Photo

Today more than 531,277 acres (215,000 hectares) of Namibian desert and savannah are protected in one of the region's largest private reserves. By day, visitors can see southern African species including zebras, springboks, kudu, hyenas and more. The most common daytime activity is a guided safari, led by locals who can help explain the dynamic landscape as well as point out the different species.

Initially designed to protect a unique ecosystem's flora and fauna, NamibRand Nature Reserve also became a sanctuary for dark skies. In 2012 it was recognised and certified as a Dark Sky Reserve; within the reserve all construction is required to comply with low-pollution lighting.

There are three main lodging options for travellers to NamibRand, including the high-end &Beyond Sossusvlei Desert Lodge. With an on-premises observatory, 12in (30cm) viewing telescope and resident astronomer, this upmarket desert accommodation is one of the best in the world for stargazing. Several tour operators also provide astronomy information as part of their guiding services through the reserve. Within the reserve itself there is also a worthwhile and wide-ranging education centre, Namib Desert Environmental Education Trust (NaDEET).

A NamibRand Nature Reserve Control Warden, Murray Tindall has learned that it doesn't have to be a special occasion to make for an enjoyable night of stargazing in the wild landscapes of the Namib desert. Tindall is responsible for ensuring that the reserve maintains its light pollution control standards, and as a result he spends time appreciating the dark skies as part of his work: 'Watching the full moon rise over the desert plains never gets old. Nor does throwing out a blanket on the ground and counting shooting stars.'

Important Info

When to visit: The warmest and wettest months are December to April. For wildlife, opt instead for the dry months of May to October.

Website:
www.namibrand.com

Lauwersmeer National Park

NETHERLANDS

As the most densely populated country in the EU, the Netherlands is one of the most light-polluted countries in the world, but it is also home to one of the great spots for seeing the night sky in Western Europe: Lauwersmeer National Park. While preserving the night sky was not originally part of the plan when Lauwersmeer was formed, the geography and location of the park made development difficult and offered the Netherlands a chance to create an enclave of darkness along its northern shore.

From left: Meervogel windmill at sunset; Wadden sea coast by Moddergat.

Lauwersmeer is a man-made lake near the Wadden Sea, now itself a Unesco World Heritage Site. It was created in 1969 to help reclaim land from an intertidal zone, and has grown into a natural refuge. The spongy, soft ground in the countryside

© Ron Buist / 500 px; © Olha Rohulya / Shutterstock

that emerged as the water drained from the region was ill-suited to human development but ideal for birds and other plants and animals. In 2003 Lauwersmeer was recognised as a national park to aid in protecting those species that live in the area, and was also designated as a Dark Sky Park in 2016. These two designations help ensure that by day and by night, Lauwersmeer will continue to be a uniquely undeveloped and wild part of the Netherlands.

By day, visitors are typically enthralled by the huge range of birds that live in Lauwersmeer National Park. Birding happens year-round but especially during the migratory spring and autumn months. Visitors can also look for species of orchids, foxes, cattle and horses that call the national park their home.

Once the sun goes down, you can look up to experience what Lauwersmeer has to offer. Though the park is less than two hours from Amsterdam by car, the city lights aren't visible here. It's possible to see the Milky Way during warm-weather months and even the aurora on especially ideal winter nights. Independent stargazers are welcome in Lauwersmeer National Park, and Staatsbosbeheer, the Dutch forestry organisation, arranges events for stargazing, viewing meteor showers and even observing or counting nocturnal species that live in the park. Simple nature shines here, with water reflecting the sky above. It makes a marvellous dark-sky escape set in the heart of Europe.

Lauwersmeer National Park recently entered into an agreement with the astronomy department at nearby Groningen University, according to Jaap Kloosterhuis, a long-serving Park Ranger at Lauwersmeer. 'Together we organise stargazing evenings where astronomy students take their telescopes and explain about stars and planets and other celestial phenomena', Kloosterhuis says. 'In the near future we hope to expand the collaboration and build an observatory.'

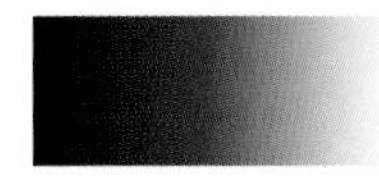

Important Info

When to visit: Spring and autumn are the best times to visit for temperate weather plus the chance to experience the migration of hundreds of bird species.

Website: *www.np-lauwersmeer.nl*

Aoraki Mackenzie Dark Sky Reserve

NEW ZEALAND

In the early days of human history, the night sky was a central character in cultural lore and belief systems. Before artificial light and electrical power, the stars were the primary show available to everyone each night, and the stories about stars, constellations and other astronomical phenomena seem almost as numerous as the stars themselves. This holds especially true for the Māori of New Zealand, who not only had complex lore about the night sky but also used the stars for navigation around New Zealand's islands.

Thanks to this history, New Zealand has long been a haven for astronomers and increasingly for astrotourists. Aoraki Mackenzie Dark Sky Reserve, in the heart of New Zealand's South Island, is one of the best places in the country to view the night sky. Comprised of Aoraki (Mt Cook) National Park and the Mackenzie Basin, the Dark Sky Reserve was certified in 2012 to continue protecting the dark skies in the area (though efforts have been underway since the 1980s).

Within Aoraki Mackenzie Dark Sky Reserve, popular daytime activities include hiking and rock climbing. Glacier-fed, turquoise Lake Tekapo and Lake Pukaki are also popular for boating. By night, visitors flock to Mt John Observatory for stargazing; they may also see the aurora on clear winter nights. Operated by the University of Canterbury in Christchurch (a three-hour drive away, so no need to worry about light pollution), this facility offers night sky tours and observation through one of its many telescopes. It's equipped with a 1.8m telescope for academic research in partnership with Japan, and also has a telescope exclusively for tourist use. Tours must be booked in advance through the tour operator Earth & Sky; trips originate from the nearby town of Tekapo.

Clockwise from top: Mt John Observatory; Aoraki (Mt Cook) looms; a visitor enjoying Tasman Glacier Lake.

Steve Butler is a dark-sky advocate who helped with the certification process for Aoraki Mackenzie Dark Sky Reserve. He remembers showing a group of guests, including politicians, the pristine night sky above Aoraki Mackenzie. 'To hear them exclaim about the vastness of the night sky was especially pleasing', Butler said. 'They said it helped them gain perspective on their normally busy and challenging lives.' With the RAS of New Zealand, Butler is working to reduce light pollution outside of reserves as well.

Important Info

When to visit
The winter months of May through August offer the best chance to see the night sky and the aurora.

Website: *https://mackenzienz.com*

Aotea (Great Barrier Island)

NEW ZEALAND

Located off the coast of New Zealand's North Island, Great Barrier Island (GBI, Aotea in the indigenous Māori language) combines low light pollution due to limited human development with the dark skies that only islands can provide. Nearly 60% of the island is managed by the New Zealand Department of Conservation, guaranteeing its protection as a natural space both by day and by night.

The island is one of only five Dark Sky Sanctuaries in the world, the highest class of certification. It's widely considered one of the best places in the southern hemisphere to view the night sky with no impact from artificial light. Visitors typically spend their days trekking and their nights camping or staying in a hut. One of the most popular walking trails, the Aotea Route, takes 2–3 days and is the only multi-day trek in the Auckland region. While travelling to Aotea (Great Barrier Island) isn't an easy task (a lack of tourist infrastructure is what keeps the skies so pristine), it's easy to enjoy the night sky once you arrive. Astronomical sights include the Southern Cross and the Magellanic Clouds, which are visible only in the southern hemisphere.

To help with orientation, local tour operator Good Heavens provides a Dark Sky Ambassadors service. Local experts point out notable features in the night sky like the Pleiades star cluster, seven stars known to Māori as 'the sisters' or Matariki, whose appearance marks the start of the Māori New Year late each May when it rises above the horizon, or else at the first new moon after Matariki rises. Increasingly it is a celebration throughout all of New Zealand. Other traditions consider Matariki a mother star, with six daughters clustered around her. While there are hundreds of stars in the cluster, these seven are visible to the naked eye. Guides provide binoculars and a telescope and offer dark-sky dinners, set under the stars.

Clockwise from top: Camping in starry New Zealand; the Southern Cross; Port Fitzroy Wharf, Aotea (Great Barrier Island).

'I have lived on Aotea (Great Barrier Island) for 30 years, and the beauty of the Milky Way never fails to amaze me', shares Izzy Fordham, the chairperson of the Aotea (GBI) Local Board. 'Being on an island that lives off the grid with no street lights or neon signs, on clear nights you can see forever. We have heard of places around the world where the Milky Way is no longer seen; we didn't want that to happen to our place.' Thanks to the island's Dark Sky Sanctuary status, future generations can continue viewing the Milky Way here.

Important Info

When to visit: Winter (June to August) offers cold, clear nights for stargazing; bring appropriate gear.

Website: *www.greatbarrier.co.nz*

Yeongyang Firefly Eco Park

SOUTH KOREA

From left: The observatory at Cheomseongdae; Seoul tower under the Milky Way.

Named for the nocturnal insect the park has helped protect, Yeongyang Firefly Eco Park is an island of darkness in a sea of light on the Korean Peninsula. As in many parts of eastern Asia, rapid development and urbanisation have reduced darkness and increased light pollution to some of the highest levels in the world. Yeongyang Firefly Eco Park is in a valley of the Wangpi River Basin Ecological Landscape Protected Zone in eastern South Korea, which has helped protect it from development. The contrast to bustling, ultra-modernised Seoul is extreme.

As it's not just fireflies who are affected by the loss of night, Yeongyang Firefly Eco Park was designated as a Dark Sky Park in 2015, the first park to receive the designation in Asia. Even before receiving its formal designation, Yeongyang Firefly Eco Park was increasingly popular with amateur

© Xiquinho Silva; © nattanai chimjanon / Alamy Stock Photo

astronomers, who recognised it as one of the few places left in South Korea where you can see the night sky clearly. The park is 2½–4½ hours by train from Busan, Daegu or Seoul – so the nearly 16 million residents of these cities, millions more in the surrounding region, and visitors can all travel to Yeongyang for an overnight trip.

The Yeongyang Firefly Eco Park facility has accommodation, an educational centre, and an observatory and planetarium. From the observatory you can use one of five telescopes to see the Milky Way, galaxy clusters, nebulae and planets – and viewing fireflies, of course, requires no telescope at all. During daytime visits, solar observation is also possible. The eco park website (in Korean) allows registration for evening observations; if you're planning a trip, it's advised that you book a session in advance.

Add in a stop at Cheomseongdae in Gyeongju, the oldest surviving astronomical observatory in Asia, and among the oldest in the world. This stone tower dates to the 7th century, and its construction seems to be based on the astronomical calendar. Even the name Cheomseongdae translates as 'stargazing platform'. The incredible remains at Cheomseongdae are a testament to the ongoing importance of astronomy throughout human history.

Important Info

When to visit: Summer (June to August) is the best time to visit for warm weather and to see the park's famous nocturnal residents. Firefly activity peaks in the summer months when days and twilight last longer, so be prepared to stay up late.

Websie:
www.yyg.go.kr/np

Albanyà

SPAIN

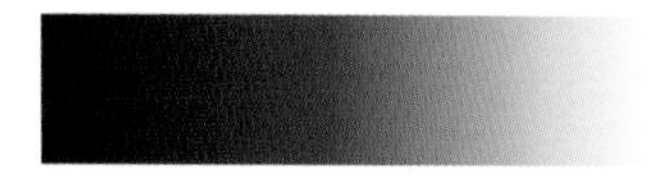

Located in the far northeast of Spain near the French border, Albanyà is typically not on most travellers' itineraries. However, this destination in Catalonia is a must for astrotourists, as Albanyà is the first officially recognised dark-sky place in Spain, certified by both the Starlight Foundation and the International Dark-Sky Association.

From left: A Spanish mountain village under the night sky; the ruins at Monestir de Sant Llorenç de Sous in Albanyà.

© marcin jucha / Alamy Stock Photo; © Josep Maria Viñolas Esteva / Creative Commons License

Astrotourism is one of the main drivers bringing visitors to Albanyà, and the entire municipality is working to establish Albanyà as a place to enjoy the night sky. The town of Albanyà itself is home to fewer than 200 people, and its rural setting ensures a natural preservation of the night sky. While you'll mostly find local tourists from elsewhere in Catalonia stargazing in Albanyà, its 2017 Dark Sky Park designation ensures more travellers are aware of the night-time wonders overhead.

For visitors to the region, Bassegoda Park is a must-visit. This accommodation and camping facility offers a variety of stargazing activities and evening programmes to help visitors better understand the sky. It helped establish Albanyà as a stargazing region and has increased its summertime astronomy programming each year to draw more travellers. Located outside Albanyà, Bassegoda Park is a good option for travellers who want to stay at the same location where they'll stargaze. By day you can explore the surrounding countryside, relax by the swimming pool and connect with fellow stargazers in the region.

The Albanyà Astronomical Observatory, opened in 2017 with a 15.7in (40cm) telescope, is another major attraction for travellers who want to appreciate the night sky. Other popular spots for stargazing in the area on your own include El Casalot and El Pla de la Bateria. Both of these are open to the public but there are no set astronomy events. As such, they may be better for travellers who know the night sky well or astrophotographers looking to escape the crowds for photo opportunities. One of the starriest, light-pollution-free skies in Catalonia, it's an ideal sky pilgrimage site.

If you don't speak Catalan or Spanish, aim to visit in the summer months when programming is also done in English, advises Astronomical Observatory Director Pere Guerra. 'Our stargazing sessions aim to be a unique sensorial experience where our team of astronomers, led by the renowned astrophotographer Juan Carlos Casado, take the visitors on an audiovisual tour through the sky, adapting the content of your stargazing experience to the current weather and time of the year.'

Important Info

When to visit: From May to September, temperatures are comfortable both day and night, with a better chance of clear skies for stargazing.

Website: *www.bassegodapark.com*

Brecon Beacons National Park

UK

In the heart of southern Wales, a microcosm of Welsh landscape allows visitors to explore rolling fields, climb mountains, experience eight millennia of human history and gaze upon some of the darkest starry skies in Britain. Brecon Beacons National Park is a short, hour-long drive from the city of Cardiff, but it feels worlds away.

Most travellers to Brecon Beacons will pass the daytime exploring the diverse landscape here, which is home to four hill and mountain ranges. Many people come to take an extended walk or trek through the countryside, where Welsh mountain ponies and sheep graze. Among these hilly formations you can find lakes and waterfalls as well as Neolithic ruins and standing stones. One of the biggest tourist attractions by day is to go underground, exploring the network of natural caves and underground mines ... no stars to be seen in the darkness, though!

By night the whole of Brecon Beacons National Park is blanketed with a starry sky, protected by the local communities and certified as a Dark Sky Reserve since 2013. Here you can stargaze from within a stone circle or on the edge of town outside the range of the low-light-pollution fixtures installed throughout the park. It's common to see the Milky Way and Andromeda galaxies on a clear night, and the northern lights have even been spotted on nights when the aurora is particularly strong.

The Cardiff Astronomical Society holds events throughout the year to encourage locals and travellers to enjoy Wales' national parks and the night sky above them. History and archaeoastronomy buffs will also want to pay a visit to the fascinating stone circles found in Powys within the park, which are considered to have astronomical alignments.

Clockwise from top: Camping in Snowdonia Dark Sky Reserve under the moon; the Milky Way over a standing stone in Wales; a Welsh Pony in the park.

Brecon Beacons is one of three national parks in Wales, and the first of two certified Dark Sky Reserves. The other is Snowdonia National Park, a waterfall- and lake-filled landscape a three-hour drive north that boasts Wales' and England's highest point. There's also a Dark Sky Park en route, Elan Valley Trust, making Wales a great destination for a stargazing road trip. Wales' third national park, Pembrokeshire Coast National Park, is a Dark Sky Discovery Site. Wales has one of the highest global percentages of its land area with dark-sky protection.

Important Info

When to visit: Outside the summer months, prepare for cooler temperatures at night (and snow in winter).

Website: *www.breconbeacons.org*

Exmoor National Park

UK

National parks can be created for a number of reasons: to safeguard unique geological formations, to protect flora and fauna, to preserve or decrease human impact on a particular area of land, or for any combination of these. Exmoor National Park was among the first national parks designated in the UK in the 1950s for all of these reasons. A mix of moorland, woodland, fields and cliff shores, Exmoor has been inhabited by humans and animals alike for over eight millennia. Wild deer and ponies wander past standing stones and Roman ruins, and you'll often spot a fellow traveller out walking the countryside. Climbing, cycling and horseback riding are other popular ways to explore Exmoor, while a visit to Dunster Castle is a must.

Exmoor has also gained recognition for what visitors can experience after the sun goes down. The oldest Dark Sky Reserve in Europe, Exmoor was certified in 2011 and is still considered the darkest place in England for stargazing. People travel from around the UK to stargaze in Exmoor, and increasingly from further abroad as well. During a festival established in 2017, visitors can experience a host of night-time activities for two weeks in late October each year. These include star parties, planetarium shows, meteor shower safaris (the Orionids peak during the festival) and astronomy talks. This festival focuses on astronomy activities, but it's possible to enjoy the night sky at Exmoor year-round.

Stop by one of the visitor centres in the towns of Dulverton, Dunster or Lynmouth to pick up a pocket guide to stargazing in Exmoor, including handy star charts to the main constellations you can see each season. It's also possible to rent a telescope from these centres to enhance your experience. Park rangers can also advise you on the top spots for stargazing in Exmoor, including some of the famous hilltops and lakes in the national park.

© Chrispo / Alamy Stock Photo; © GAPS / Getty Images; © incamerastock / Alamy Stock Photo

Clockwise from top: Milky Way from Valley of the Rocks in Exmoor; a sheep in the purple heather at the park; approaching the landscape of Valley of the Rocks.

Ben Totterdell serves as the Interpretation and Education Manager at Exmoor National Park, working to educate the public on its night skies. While the astronomy nights are a great way to experience the park, Totterdell encourages visitors to attend even when events are not scheduled: 'Although we run events where people can use telescopes to explore space beyond the range of the naked eye, for me the simple pleasure is being able to look up and marvel at a clear, starry Exmoor sky.'

Important Info

When to visit: Exmoor has temperate weather for stargazing all year. Clouds and rainy weather are common in the cooler months of October through March, interfering with night sky views.

Website: *www.exmoor-nationalpark.gov.uk*

Galloway Forest Park

UK

From left: Male sparrowhawk in the park; heather abounds in Galloway's hills.

As is the case with many lands protected during the 20th century, Galloway Forest Park did not initially set out to preserve the night skies above the picturesque mountains, hills, valleys and lakes in its confines. The park, managed by Forestry Commission Scotland, was instead established in 1947 to protect 299 sq miles (774 sq km) of countryside in southern Scotland for visitors who wanted to go hiking, mountain biking or even ice climbing.

Over time it became apparent that Galloway Forest Park offered something special for visitors at night too, and in 2009 it became the first designated Dark Sky Park in Great

© smharperphotography / Alamy Stock Photo; © travellinglight / Alamy Stock Photo

Britain. Since then a core section of the park has been protected from light pollution to an even greater degree, with no permanent illumination or light sources. Within the heart of Britain's largest forest park, you can see the night sky at its darkest.

Galloway Forest Park is ideal for travellers who want the flexibility to experience the park and view the night sky at their leisure. Three visitor centres provide information on activities within the park as well as dark-sky viewing spots. The visitor centre at Clatteringshaws is considered the best base for stargazing, as it is located close to the dark core area of the park. Camping is allowed within Galloway Forest Park, so you can even choose a place in the park to spend the night.

Augment your trip with a visit to the Scottish Dark Sky Observatory, situated near the northern border of the forest park. Attend an evening stargazing event to view the sky through the 20in (51cm) Corrected Dall-Kirkham telescope and 14in (36cm) Schmidt-Cassegrain telescope. The observatory also has a planetarium for daytime visitors. Be sure to check the online observatory events calendar as pre-bookings are required to attend any event, and the facility is only open on the nights noted on the calendar. Another option is the nearby Selkirk Arms Hotel, which has stargazing weekends with professional astronomers.

If you're unable to travel to Scotland to experience the dark skies over Galloway Forest Park, you can still enjoy the view using enhanced reality to experience it remotely. VisitScotland, the official tourism agency for Scotland, has created a virtual reality app that includes a Galloway Forest Park dark-sky experience.

Important Info

When to visit: As with other dark-sky locations in the UK, summer (May to August) is the best season to visit for warm weather and clear skies. Summer will also mean shorter nights, so be prepared to stay up until midnight or later for the darkest skies.

Website: *https://scotland.forestry.gov.uk/forest-parks*

Northumberland International Dark Sky Park

UK

Hugging Hadrian's Wall and the border between England and Scotland is the largest protected area of dark skies in Europe. The Northumberland International Dark Sky Park was designated in 2013 as one of the best places to see the night sky in the UK. Here you can explore nearly 615 sq miles (1593 sq km) of English wilderness, comprising Northumberland National Park and the adjacent Kielder Water & Forest Park, by day and by night. Northumberland International Dark Sky Park was the first of its kind to combine two separate parks under a single dark-sky designation.

Northumberland International Dark Sky Park is well situated to allow residents and travellers to experience the night sky, accessible from Edinburgh, Newcastle and Carlisle. It's a short 90-minute drive from the Scottish capital to go stargazing in the park or visit the Kielder Observatory for one of the astronomy talks or night sky events. There are also communities throughout the park that comply with light pollution ordinances, so you can spend the night in the area too.

By day, visitors can enjoy walking, hiking or cycling on designated trails throughout the parks. History lovers will also appreciate the area's castles, fortified homes, chapels, towers, and stone walls. Named in honour of the emperor who ordered it built, Hadrian's Wall was one of Rome's greatest engineering projects. This enormous 73-mile-long wall (117km) was built between AD 122 and 128 to separate the Romans and Scottish Picts. Today, the awe-inspiring sections that remain testify to Roman ambition and tenacity. When completed, the mammoth structure ran across the island's narrow neck, from the Solway Firth in the west almost to the mouth of the Tyne in the east. For every Roman mile (0.95 miles/1.5km)

The northern lights and the Milky Way glow over Coquet Lighthouse, Northumberland.

© Ollie Taylor / Alamy Stock Photo

there was a gateway guarded by a small fort (milecastle), and between each milecastle were two observation turrets. Back in the days of the Romans and after they left, the skies beamed down in their full glory each night on the milecastles.

Some other popular stops in the region include Dally Castle, Biddlestone Chapel and Black Middens Bastle House. Most of these historic sites are open to the public, offering visitors a sense of English life over the centuries. There are ample opportunities for independent stargazing, and it's regularly possible to see the Milky Way on a clear night.

Duncan Wise at the National Park Authority says, 'Northumberland International Dark Sky Park has the darkest, most pristine skies of any county in England. Compared to towns and cities, where you may struggle to see more than 5–10 stars, when you are up on the central part of Hadrian's Wall, for example, you can see more than 2000 stars and the Milky Way.' Some of the top spots for stargazing here include Cawfields, the dark sky community of Stonebaugh and the ruins of 12th-century Harbottle Castle.

Important Info

When to visit: Weather is typically English. Expect rainier and cloudier weather from September to April. Visit in summer for warmth and the best chance of clear skies.

Website: *www.northumberlandnationalpark.org.uk*

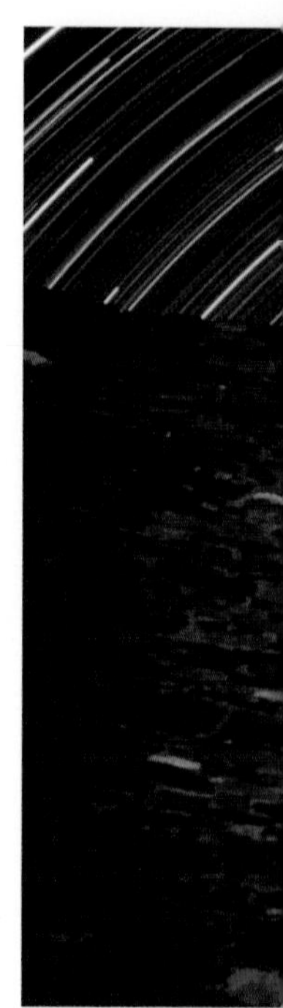

Chaco Culture National Historical Park

USA

From left: Pueblo Bonito, the remains of an Ancestral Puebloans great house; star trails above Chaco Culture.

The night sky has played a significant role in various human cultures throughout millennia, including those of the American Southwest. The Pueblo peoples who lived in the Four Corners region (the area where the borders of Arizona, Colorado, New Mexico and Utah meet) were known for their architecture, engineering and culture; the remains of the large stone structures they built in Chaco Canyon are widely considered some of the most impressive from that era still standing in the region. Additionally,

© Inge Johnsson / Alamy Stock Photo; © age fotostock / Alamy Stock Photo

archaeoastronomers posit that many of the buildings here were built in alignment with celestial events.

Today the Chaco Canyon site in New Mexico is preserved as part of the US national park system, and it was designated as an International Dark Sky Park in 2013. The light pollution controls established to protect the night sky above Chaco Canyon mean that visitors can experience the night skies as the Ancestral Puebloans did centuries ago.

Since the late 1990s the National Park Service has worked to establish the Chaco Night Sky Program, encouraging visits to the Chaco Observatory. There are opportunities to attend guided educational astronomy events, have questions answered by astronomers and volunteers, and participate in biannual star parties in partnership with the Albuquerque Astronomical Society. The park also offers unique events on the solstices (June and December) and equinoxes (March and September) where visitors can learn about the alignment of buildings within Chaco Canyon and how the seasons and the passage of time were marked by ancient people.

By day, popular activities include hiking and biking, as well as visiting the ruins. There are still 19 Pueblo tribes resident in New Mexico, with each tribe a sovereign nation with its own distinctive culture. Nearly all still make traditional Pueblo bread, baked in *hornos* (adobe-built ovens).

Archaeoastronomy, the so-called 'science of stars and stones,' is the interdisciplinary study of how ancient cultures used the night sky as part of culture and society – including in construction. Sites like Stonehenge in England and Chichén Itzá in Mexico are among the locations of interest to archaeoastronomers, since they seem to be aligned with celestial events such as equinoxes and solstices. Archaeoastronomers use material remains to examine how ancient cultures related to phenomena in the sky.

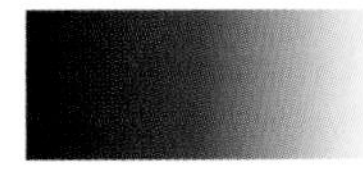

Important Info

When to visit: Be prepared for desert weather. Summers are hot by day, and winters are cool by night. The transition seasons of spring (March to May) and autumn (September to November) are ideal for visits.

Website:
www.nps.gov/chcu

Cherry Springs State Park

USA

In the eastern US it can be hard to find a view of the dark sky. Massive urban and suburban development have slowly reduced the opportunities for residents and travellers to marvel at the stars. Cherry Springs State Park, in the northern part of central Pennsylvania, is one of the remaining spots where you can see the night sky, and its 2008 designation marks it as one of the world's earliest Dark Sky Parks. Amateur observers across the Eastern seaboard have made this a premier destination for bringing their telescopes and peering up at the sky.

The primary attraction for visitors to the relatively small Cherry Springs State Park is the stars overhead. The park is surrounded by the Susquehannock State Forest, which serves to insulate it from light pollution in the region. The most popular spot for stargazing is the Astronomy Field, where you can enjoy a 360-degree view of the night sky from horizon to horizon. Stargazers regularly see the Milky Way, planets, meteors and even nebulae when the conditions are ideal. If you have a telescope or binoculars, the skies reveal even more wonders.

Star parties are held biannually at Cherry Springs State Park, typically

The Milky Way reveals itself in the night sky at Cherry Springs State Park.

© Elizabeth Creason / Alamy Stock Photo

in June and September, when astronomers from the surrounding region flock to the park, and the public is welcome to join them. With between 60 and 85 days of perfect stargazing conditions each year, you can visit almost anytime. While winter weather will be chillier, it also offers the chance to see constellations and even the aurora.

Visitors can camp at designated spots within the park or stay in the nearby community of Coudersport or surrounding Potter County. By day you can go hiking and spot wildlife including elk and bald eagles. The real allure for visitors to this corner of Pennsylvania, otherwise unremarked by most but precious to devoted stargazers, is the skies above.

Scott Morgan works as Assistant Park Manager for Cherry Springs State Park, hoping for visitors to gain appreciation for the dark sky by attending astronomy events. 'My most moving experience is the crowd reaction to the Perseid meteor showers', he shares. 'During the peak it is not unheard of to witness over a hundred meteors per hour. The crowd reaction is often hundreds of oohs and aahs. It is quite the event to witness in person.'

Important Info

When to visit: Expect more crowds during the warm summer months (May to September). If you're planning to stay overnight in the park, register in advance to reserve your spot. Pack layers for winter travel.

Website:
https://cherryspringsstatepark.com

Cosmic Campground

USA

With a name like this, it will likely come as no surprise that Cosmic Campground is one of the darkest places in the entire US. In fact it's one of only a few places certified as a Dark Sky Sanctuary in the world, all in remote locations (the other one of these located in the US is the Rainbow Bridge National Monument in Utah). While small by the standards of some dark-sky locations, Cosmic Campground has 3.5 acres (1.4 hectares) of unsullied night darkness in the heart of the Gila Wilderness and Blue Range Wilderness in western New Mexico. Far less well known than New Mexico parks such as White Sands National Monument and Carlsbad Caverns National Park, Cosmic Campground is nonetheless a pilgrimage site for astrotourists.

Travellers to Cosmic Campground are there for one thing: darkness. With the nearest significant source of artificial light over 40 miles (64km) away in Arizona, there are few places darker. You won't find luxury accommodations while stargazing in Cosmic Campground; the most you'll have is basic camping infrastructure (think pit toilets and no electric hookups). Instead, bring a telescope and set up in one of the designated observation pads outside the camping area, where you'll have a view of the unobstructed night sky from horizon to horizon in every direction.

There are some rules if you choose to spend a night in Cosmic Campground. Its status as a Dark Sky Sanctuary means that it's every visitor's responsibility to preserve the darkness. The rules include arriving before sunset so you don't use car headlights, covering all other light sources with red filters, and not starting campfires near the observation areas. As Cosmic Campground is one of the darkest places you'll likely ever visit, respect for the rules helps ensure an unforgettable night of stargazing for you and any other observers in the area. By day, visitors can explore the Gila National Forest by hiking or horseback riding, or visit the Gila Cliff Dwellings National Monument, a three-hour drive from Cosmic Campground.

Clockwise from top: Gila Cliff Dwellings National Monument; a Southwest star party; zodiacal light in the Gila Wilderness.

Patricia Ann Grauer is a Friend of the Cosmic Campground and worked to help establish it as the first Dark Sky Sanctuary in the northern hemisphere. 'I suggest being there near the new moon, before sunset if possible', she recommends. 'As the sun sets, the natural light subsides. The natural night sky reveals itself as planets, stars, constellations, clusters and the Milky Way begin to light the sky. Without the pollution of artificial light, your night vision allows you to see people, objects and the ground, often without a red flashlight.'

Important Info

When to visit: Summers are hot by day and winters are cold at night, making spring and autumn ideal.

Website: *www.fs.usda.gov/recarea/gila/recarea/?recid=82479*

Craters of the Moon National Monument and Preserve

USA

Nestled in southern Idaho's Snake River Plain, Craters of the Moon National Monument and Preserve is one of those special places on Earth where you feel like you've stepped onto another planet ... or in this case, a moon!

Craters of the Moon first gained national attention in a 1924 issue of National Geographic, when an early visitor dubbed its geological features 'craters of the moon' because they appeared similar to those of our closest celestial neighbour. The site was protected as a national monument that same year, and travellers still come to see the weird basalt and lava formations that cover the region. Unlike the moon's craters, which are caused by interstellar impacts, Craters of the Moon is a volcanic site, with lava tubes, cinder cones and massive solidified lava flows. Despite its otherworldly appearance, it is a distinctly terrestrial landscape, evidence of our own planet's tumultuous geological history.

The unusual terrain and high elevation give you the sense that you're off-world, and by night you might just believe it. During warm-weather months Craters of the Moon is open for camping at several campgrounds as well as in the backcountry, if you're properly equipped and really want to escape the crowds. Astrophotographers will take advantage of the lava and rock formations to compose striking shots.

Travellers have visited Craters of the Moon for decades, hiking in the summer and cross-country skiing in the winter. In 1969 NASA even used the site as training grounds for moon-bound astronauts so they might be better prepared for their missions. Pilots rather than geologists, Craters of the Moon was the perfect place for these budding space explorers to practice collecting rock samples

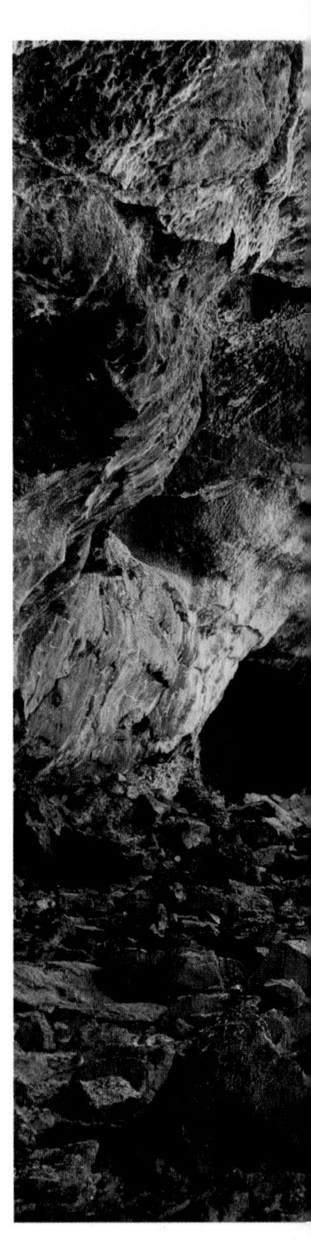

Stairs descending into lava tube cave Indian Tunnel at Craters of the Moon in Arco, Idaho.

© Anna Gorin / Getty Images

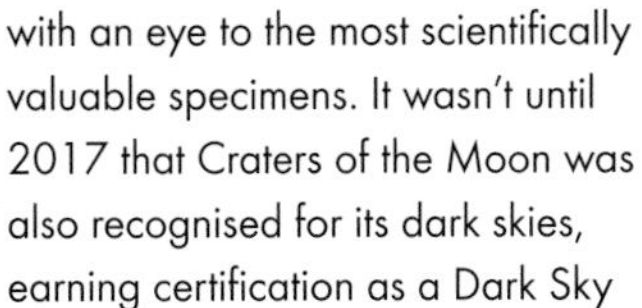

with an eye to the most scientifically valuable specimens. It wasn't until 2017 that Craters of the Moon was also recognised for its dark skies, earning certification as a Dark Sky Park from the International Dark-Sky Association. It sits next to one of the largest areas of dark skies in the US, itself recognised in 2017 as the Central Idaho Dark Sky Sanctuary.

Wade Vagias works at Craters of the Moon and was part of the team that petitioned for its Dark Sky Park status. 'I want visitors to Craters of the Moon to have the chance to experience and to be inspired by a night sky as dark and pure as it was before the advent of modern lighting', he says. 'One night I realised I had never been to a place as dark as Craters of the Moon was that night – it was a new moon, and the stars were breathtaking. It's a memory I'll never forget and one I hope other park visitors will have too.'

Important Info

When to visit: Summer is the best season to visit, but brings crowds. In the shoulder seasons (spring and fall), be prepared for colder nights.

Website:
www.nps.gov/crmo

Grand Canyon National Park

USA

When discussing the top stargazing locations around the world, it's impossible to leave out Arizona, one of the best preserved and most recognised destinations for dark-sky tourism on Earth. The International Dark-Sky Association was founded in Tucson, and Arizona has more designated dark-sky places than any other state in the US. The crown jewel among them is Grand Canyon National Park, which was certified as a Dark Sky Park in 2016.

The Grand Canyon is one of the great natural wonders of the world, descending over 6000ft (2000m) through millennia of rock layers in the northwest Arizona desert. The national park that protects the canyon is a large 1901 sq miles (4924 sq km) and receives over six million visitors per year.

Most people come to Grand Canyon National Park to admire the geological features or to hike, climb or go rafting in the canyon on the Colorado River that formed it. Stargazing is also an increasingly powerful draw for visitors who want to see the night sky as the indigenous Navajo, Hopi and Havasupai peoples who live in the area have for centuries.

Each June the National Park Service holds a week-long star party to educate visitors about the importance of dark skies for flora and fauna – including humans – and the impacts of light pollution. Events include lectures and talks by prominent astronomers, park rangers, filmmakers and photographers, as well as solar and night viewing through telescopes set up on both the North and South Rims. If you're unable to visit in June, park rangers also host evening talks and walks from the visitor centres through the autumn months. Whenever you visit, check to see what programming is scheduled during your stay.

Clockwise from top: The sky over the Desert View Watchtower; sunrise over Yavapai Point; stargazing at the South Rim.

© Sapna Reddy Photography / Getty Images; © Gary Henderson/500px; © M.Quinn / NPS

If you want to avoid the crowds while stargazing, neighbouring Grand Canyon–Parashant National Monument is a great alternative. Parashant National Monument actually received its Dark Sky Park designation in 2014, earlier than Grand Canyon National Park! The national monument is considered one of the most remote national park units and lacks paved roads or visitor services, making it truly off the beaten path and ideal for observing.

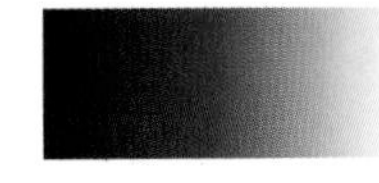

Important Info

When to visit: Spring (March to May) and autumn (September to November) are perfect for visiting the Grand Canyon. The Grand Canyon does receive snow during the winter months, so pack layers.

Website:
www.nps.gov/grca

Headlands International Dark Sky Park

USA

By the northernmost point of Michigan's Lower Peninsula, a 600-acre (243-hectare) parcel of land sticks out into the waters between Lake Michigan and Lake Huron. Primarily covered in old-growth forest, Headlands International Dark Sky Park was among the earliest to receive this designation, back in 2011. Since then, Headlands has increased its offerings for tourists seeking some of the darkest skies in the country.

Travellers who visit Headlands are undoubtedly there for the night skies, and for good reason. The Milky Way is regularly observable at Headlands on clear nights, and

The Mackinac Bridge spans the Straits of Mackinac near Headlands International Dark Sky Park.

© David R. Frazier Photolibrary, Inc. / Alamy Stock Photo

the aurora borealis is even spotted occasionally, especially during the darkest winter months. An event centre and observatory area provide a focus for astronomical events, which include free public viewings and special viewings for events like meteor showers. Park staff offer nature hikes, lantern walks and events to mark the solstices; the Northern Michigan Astronomy Club also hosts star parties throughout the year. Almost any weekend, you can visit Headlands and there will be an event or viewing of some sort going on – assuming the skies are clear!

By day, visitors can enjoy walking along the 5 miles (8km) of well-marked (and often paved) trails. One popular option is the Dark Sky Discovery Trail, a 1-mile (1.6-km) trail that educates walkers on the cultural and historical significance of the night sky for the indigenous peoples of Michigan. There's another trail marked with signs for the planets in our solar system, teaching about the mythology and discovery of each. You can't go wrong by visiting the public access beach either, with its small bay on Lake Michigan.

Although Headlands is open 24 hours a day, every day, there are no camping facilities in the park. Instead, you can stay in nearby communities like Mackinaw City, with its ferry to scenic Mackinac Island (also known for its famous fudge). Sleeping Bear Dunes National Lakeshore to the south is worth a stopover too.

Headlands Park Manager Shelly House was part of the team that worked to secure its Dark Sky Park designation. In her many years helping visitors experience the night sky, she says 'Some of my fondest memories include the "aww" when guests see the Milky Way for the first time. Probably the most profound experience was being with a visitor from LA and experiencing stars for their very first time.'

Important Info

When to visit: Summers are the best for warm weather and comfortable temperatures at night, but you'll need to stay up later for total darkness (and be sure to bring bug spray). Winters can be bitterly cold but also offer more time to stargaze due to shorter days.

Website: *www.midarkskypark.org*

Natural Bridges National Monument

USA

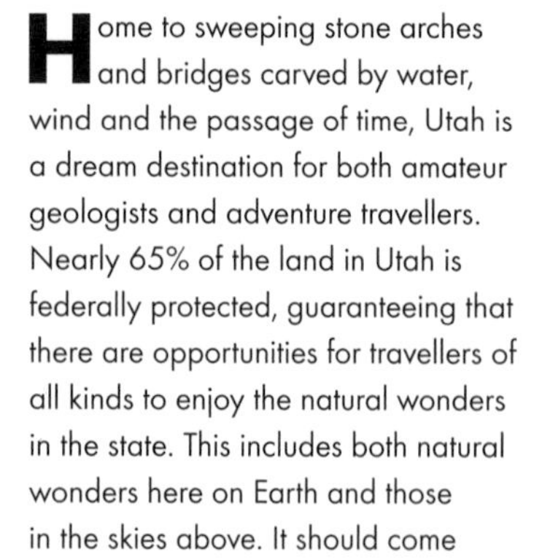

Home to sweeping stone arches and bridges carved by water, wind and the passage of time, Utah is a dream destination for both amateur geologists and adventure travellers. Nearly 65% of the land in Utah is federally protected, guaranteeing that there are opportunities for travellers of all kinds to enjoy the natural wonders in the state. This includes both natural wonders here on Earth and those in the skies above. It should come

© Yvonne Baur / Shutterstock

as no surprise, then, that Utah's first nationally protected land, Natural Bridges National Monument, was designated the world's first Dark Sky Park in 2007. Natural Bridges could well be considered one of those bucket-list destinations for those who love stargazing.

By day you should check out the three famous natural bridges for which the space is named. They are Kachina, Owachomo and Sipapu, names derived from the native Hopi people who once called this area home. These sandstone bridges were formed by erosion over thousands of years, and all along, the stars have whirled overhead each night. In the 20th century, steps to protect the natural bridges resulted in dark-sky preservation too. Now visitors can enjoy hiking and camping in the national monument, together with stargazing sessions admiring the heavens once the sun goes down.

The National Park Service works to reduce light pollution in Natural Bridges through low-energy, low-impact bulbs and motion-activated lighting in public-use areas (like restrooms). Park rangers also offer evening astronomy programmes during the summer months to encourage visitors to appreciate the night sky. Nearly 100,000 people visit Natural Bridges National Monument each year, and efforts are ongoing to ensure that as this number grows, the impact on both natural terrain and the night sky is minimised.

Photographers will take special note that Natural Bridges is one of the top spots for astrophotography in the world. The natural rock bridges, so iconically Western, form a compelling foreground against which the easily visible Milky Way can be shot for a truly stunning image.

The Milky Way above Owachomo Bridge.

If Natural Bridges National Monument inspires you to continue exploring this part of the world, consider planning a trip to nearby Rainbow Bridge National Monument. This designated Dark Sky Sanctuary, the first NPS location to secure the designation, is not easy to reach. It can be accessed only by boat across Lake Powell or via a 14-mile (23-km) hike; permit from the Navajo Nation required. Rainbow Bridge is considered one of the darkest places in the world.

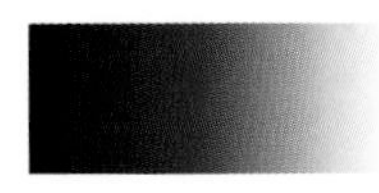

Important Info

When to visit: Summer days are hot (95°F/35°C or more) and winter nights are cold (32°F/0°C). Visit in spring or autumn for temperate weather during the day and at night. In spring there are wildflower blooms, and in autumn there is some fall foliage to enjoy.

Website:
www.nps.gov/nabr

Astronomy in Action

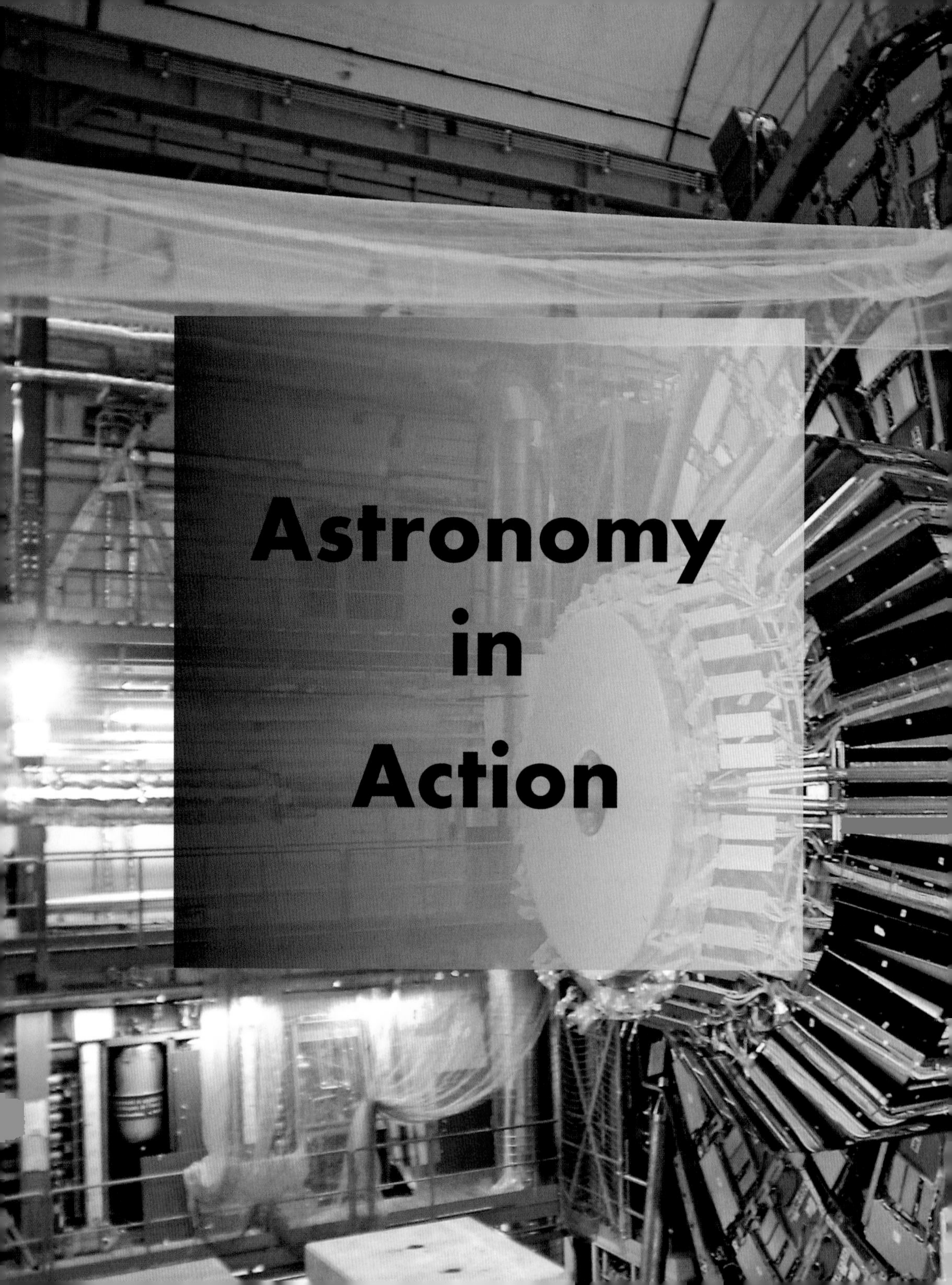

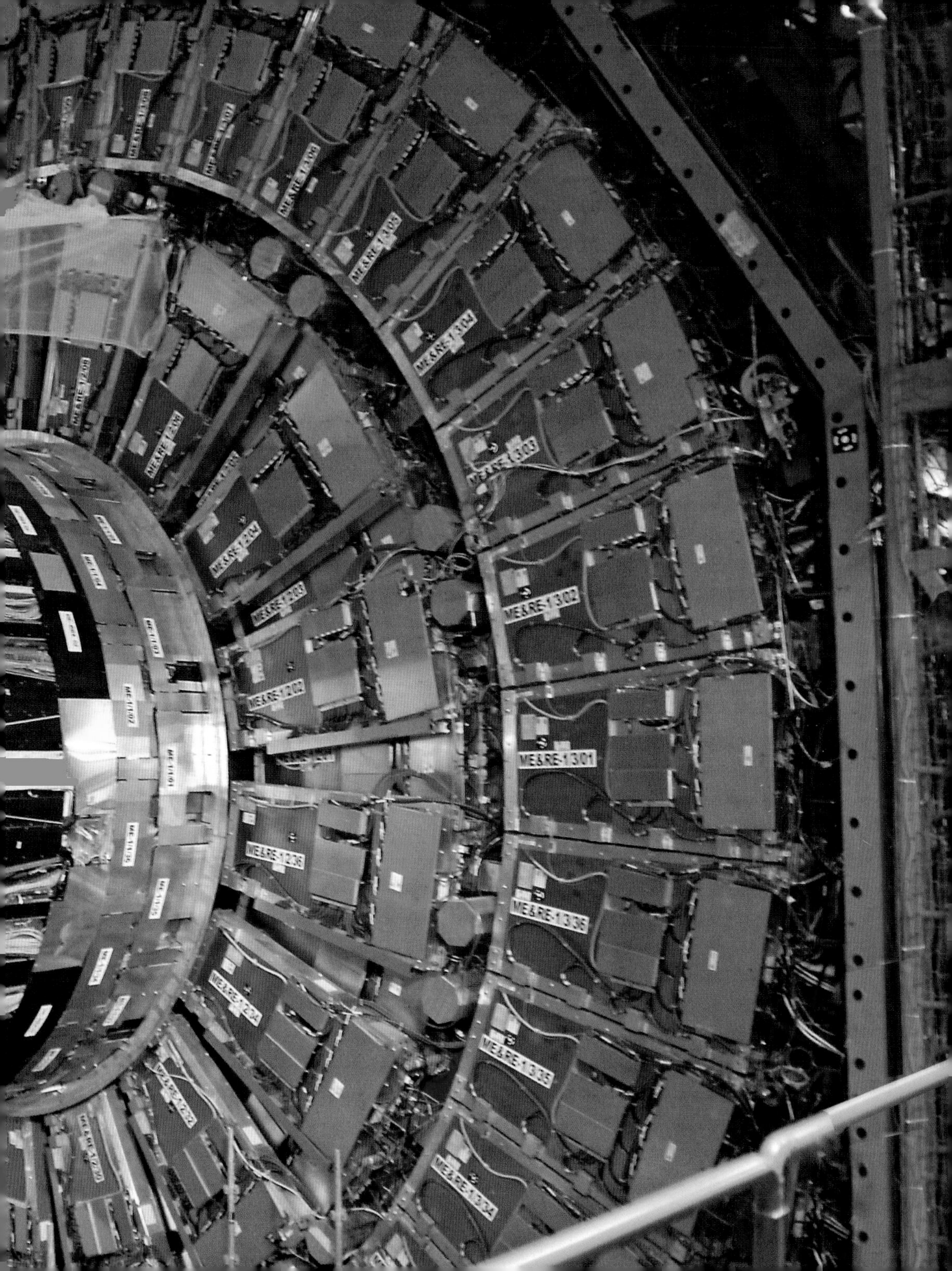

Seeing a dark night sky is one of the great wonders of the natural world, and once you've had the experience, you may be inspired to learn more about the tapestry above or to get a closer view of the action. Even with the help of the internet and access to an amateur telescope, there is only so much you can learn about the night sky and the countless interesting things on view above without getting additional guidance. This is where observatories and other research facilities come in; staff and researchers at these locations can teach you more about astronomy and stargazing, and visiting them is a chance to get up-close and personal with some of the research sites that are revolutionising our knowledge of the cosmos.

The locations profiled here are astronomical sites worth travelling to, including observatories and other research facilities studying the great mysteries of the universe, from the smallest particles to the largest galaxies. They're often in places with excellent night sky conditions, so you can also go stargazing on your own. Many of them are open to the public, though most offer limited hours or daytime-only visits to allow researchers to conduct their work without interruption or light pollution caused by visitors.

Around the world, astronomers work in a variety of scientific fields, combining physics, chemistry, biology and other sciences to advance human knowledge of space. Many astronomers work in observatories; these can be either ground-based as at Mauna Kea or the South African Astronomical Observatory, or space-based, using space telescopes like Hubble or the Spitzer Space Telescope operated by NASA Jet Propulsion Lab. Astronomers also gain knowledge by looking across the spectrum of light frequency, using optical telescopes like those at McDonald Observatory or radio telescopes like those at Arecibo Observatory to give us a complete picture of our galaxy and beyond. The instruments used to gather this data comb across the entire electromagnetic spectrum. Visible light rays (what we see when we view the stars with the naked eye) are actually only a small part of this spectrum; radio waves, infrared, ultraviolet, X-rays and gamma rays are also examined for the information they contain about far-off objects. Ground-based observatories like the ones featured here often focus on radio waves, which can be captured by antennas, and visible and infrared light, which are gathered at large optical telescopes. The technique of spectroscopy can help parse the information encoded in these rays. Other electromagnetic waves such as X-rays are best received in space and are monitored by telescopes in orbit (think of the Hubble telescope), where Earth's atmosphere doesn't get in the way.

Not all of the work being done in astronomy, astrophysics and space science is observation-based. In addition to visiting cutting-edge and historic telescopes, astrotourists can delve into the secrets of the universe at research sites devoted to rocketeering at NASA Marshall Space Flight Center in Huntsville, Alabama (also known as Rocket City, USA) or probe the furthest reaches of astrophysics at CERN, where state-of-the-art technology recreates the conditions of the Big Bang.

Visiting these sites will give you the chance to interact with the professional astronomers trying to better understand space and our place in it. You'll have the chance to speak with the scientists pushing our knowledge forward, whether it's understanding the building blocks of the universe or the geology of our solar system neighbours like the moon and Mars. In most cases, you'll also get to learn about and look through the same instruments they use to study the cosmos. You'll certainly come away with a richer knowledge of deep space and appreciation for our home planet.

Elqui Valley

CHILE

Chile, like Hawai'i and the Canary Islands, is considered one of the top astronomical observation destinations in the world. In addition to San Pedro de Atacama high in the Andes (p114), the Elqui Valley north of Santiago offers its own wealth of opportunities for astrotourists. Near the southern edge of the vast Atacama Desert, mountain rivers wind their way through a sun-baked landscape of vine-covered hillsides and serene villages, with Andean peaks looming on the distant horizon. This is the Elqui Valley, famed for its rain-free climate (over 320 days of sunshine) and remarkably clear skies. This zone of the Coquimbo Region plays host to both private research observatories on local peaks Cerro Tololo and Cerro Pachón – such as the Association of Universities for Research in Astronomy Observatory (AURA-O) and its facilities – and public observatories. The area surrounding AURA Observatory was designated as the Gabriela Mistral Dark Sky Sanctuary in 2015. This means it is one of the darkest areas of protected skies in the world, and a prime spot for stargazing even though the observatory is not open for public visits at night.

AURA Observatory will also be home to the Large Synoptic Survey Telescope (LSST), set to begin operations in the early 2020s. This telescope will sample the entire night sky with its massive field of vision.

While the research telescopes at AURA-O are not open to the public, you can visit some of the other regional astronomical attractions. Mamalluca Observatory is partly funded by the Inter-American Observatory and aims to educate visitors on both indigenous and modern astronomy, while the Pangue Observatory is run by a retired astronomer. Elqui Domos has geodesic domes that you can sleep in after an evening of gazing skyward; Cosmo Elqui similarly offers a hostel with an observatory on-site. Accommodations focusing on astrotourism are abundant throughout the region.

Clockwise from top: The night sky over the Elqui Valley; the Milky Way in the Elqui Valley; an observatory in Vicuña.

AURA-O is home to many telescopes with companions elsewhere in the world: Gemini Observatory's telescope here has a companion at Mauna Kea in Hawai'i, and the Victor M Blanco Telescope at Cerro Tololo Inter-American Observatory has a companion at Kitt Peak in Arizona. Having two telescopes in different regions allows researchers a completer view of the sky than otherwise possible, giving researchers greater knowledge about observed objects.

Important info

When to visit: Travellers may want to avoid winter (June–August) and focus on the spring and fall seasons.

Location: The Elqui Valley is south of the Atacama Desert and north of Santiago.

Website: *https://chile.travel/en/what-to-do/astrotourism/nighttime-visits*

San Pedro de Atacama

CHILE

In the Chilean desert near the point where the borders of Chile, Argentina and Bolivia meet, the small town of San Pedro de Atacama is a prime destination for travellers who want to experience life high in the Andes with the amazing stargazing opportunities the desert climate provides. Though San Pedro de Atacama is home to only 5000 people, it is a top tourism destination in Chile for both adventure and astronomy tourism. You can try sandboarding or rock climbing in the morning, explore archaeological ruins by day, and see a sky full of stars once the sun goes down. This ideal stargazing location has drawn several internationally supported observatories to the region, known as the Chajnantor Science Reserve due to their proliferation. High elevation and low humidity serve to reduce the amount of signal interference.

Of the telescopes and observatories surrounding San Pedro de Atacama, only some are open to the public. The 66-radio telescope interferometer Atacama Large Millimeter/submillimeter Array (ALMA) is open for visits on most Saturday and Sunday mornings, and advance tickets are required to see the most expensive ground-based telescope in existence yourself; note it's not possible to visit ALMA at night. The Ahlarkapin Observatory is a private observatory run by local guides, with night tours that last between 90 minutes and two hours. Other well-known observatories in the region, like the Atacama Cosmology Telescope and Simons Array, are not open to the public.

Additional viewing options include stargazing tours that operate out of San Pedro de Atacama. Several of these use local guides to take you to an area of reduced light pollution, teach you astronomy and allow you to stargaze; some offer telescope viewings as part of the tour. The Atacama Lodge even provides the convenience of telescope rental at its accommodations, a growing perk of

A vibrant red and orange sunrise at Las Campanas Observatory.

© Alberto Ghizzi Panizza / 500 px

hotels here aiming for the stargazing crowd. Tours also go to popular sites like the Valle de la Luna (Valley of the Moon) for longer stargazing-focused tours. Chile is perhaps the biggest country for astrotourism in the world (though competitors are raising their heads), so expect a variety of offerings and no lack of choice. The stars are big business here.

It's a seven-hour drive across the Bolivian border to visit the Salar de Uyuni (see p42), or salt flats, but the trip is well worth it if you're in the area. The Salar de Uyuni is another prime stargazing site in the Andes, and multi-day tour options allow you to stay in the desert overnight and enjoy stargazing in between days of adventure and exploration.

Important Info

Location: San Pedro de Atacama, Antofagasta Region, Chile.

Hours: Variable, depending on tour.

Admission: Variable, depending on tour.

Website: *https://alarkapin.cl, www.almaobservatory.org*

Pic du Midi

FRANCE

Located high atop the Pyrenees in southern France, the Pic du Midi Observatory towers over the surrounding countryside. Construction began in 1878, and it remains the grandest observing site in continental Europe. A museum and a restaurant grace the summit, and overnight guests enjoy special cultural programming that complements their access to the scientific facilities.

To visit the Pic du Midi, you must take a 15-minute cable-car ride from the town of La Mongie. During the ascent, watch the Pyrenees unfold until you reach the Pic du Midi summit at 9439ft (2877m) in elevation. In addition to walking around the observatories, you can stroll the perimeter of the Pic du Midi to admire the view in all directions or step out onto 'the Pontoon,' a suspended

© Photononstop / Alamy Stock Photo

The Pic du Midi has been designated a Dark Sky Reserve by the International Dark-Sky Association. As with other dark-sky locations, this means it is a great place to see the stars with limited light pollution or the need for telescopes. Summer is an ideal time to plan an evening visit to the Pic du Midi, as you can enjoy panoramic views of the Milky Way from the top of the Pyrenees.

walkway above the mountain. For the astronomically inclined visitor, there's a planetarium with regular shows and a museum with information about the history of the Pic du Midi Observatory. A new experimentation area explains current astronomical research happening here, including the study of the sun, cosmic rays and Earth's atmosphere. There are also evening programmes on select nights (usually weekly) throughout the year. These allow you to see the sunset from atop the Pic du Midi, enjoy dinner and do a bit of stargazing before your descent via cable car.

Every summer the Pic du Midi hosts a series of special night events called A Magical Night at the Summit of the Pic du Midi. This overnight excursion includes a cable-car ride to the summit of the Pic du Midi, sunset cocktails, dinner, guided stargazing and observing through telescopes not otherwise open to the public, sunrise viewing and a VIP tour of the facilities. Once a year the Pic du Midi offers the event in English, so be sure to check the website for dates and details; the views make it an extra special treat for stargazers and armchair astronomers.

From left: Pic du Midi de Bigorre Observatory (2,877m); the cable car to the Pic du Midi.

Important Info

Location: Rue Pierre Lamy de la Chapelle, La Mongie.

Hours: Daytime access daily 9am-4pm most months (the facility is closed most of April and all of May). Evening access varies.

Admission: Daytime, adult/child €40/24.50; evening, adult/child €129/99; rooms from €339

Website: *http://picdumidi.com*

Arcetri Astrophysical Observatory

ITALY

Italian astronomer, physicist and engineer Galileo Galilei spent the last years of his life in Arcetri, a hilly region south of Florence, where he was under house arrest for heresy as a result of his support of the Copernican theory of heliocentrism. Now, near the villa where Galileo died, visitors can pay homage to the father of modern astronomy at the Arcetri Astrophysical Observatory and learn about the work conducted here. Less than 30 minutes by car from central Florence, Arcetri is an easy day trip for travellers.

The Arcetri Astrophysical Observatory is an active facility that conducts research and observation, and takes part in instrument design and construction for observatories around the globe. Some of the famous observatories that the team at Arcetri has assisted include the Very Large Telescope and the Atacama Large Millimeter/submillimeter Array in Chile and the Gaia space observatory, launched in 2013 to map our galaxy.

You can visit Arcetri Astrophysical Observatory by day or by night on most weekdays throughout the year. During the day, visitors can participate in solar observation, looking at the sun's photosphere and chromosphere and observing cosmic rays through several of the observatory's instruments and telescopes. At night, telescopes are set up for celestial viewing of objects of interest above. Astronomers and staff from the National Institute for Astrophysics (INAF) also provide astronomical tutorials, and in the event of clouds they show videos or schedule an astronomy lecture.

In addition to tours, the observatory also organises a lecture series for hardcore astronomy buffs who want to learn about more complex astronomy topics. These typically occur weekly on Thursdays; it's best to check in advance if the seminar will be offered in Italian or English. It's a unique chance to learn about astronomy on ground once trod by one of its leading figures.

© Javen / Shutterstock

Pay homage to Galileo at other spots in and near Florence, including the Villa Galileo in the Arcetri hills, where Galileo spent the last 10 years of his life, and the Museo Galileo, a museum that honours Galileo's scientific contributions. The museum houses the renowned Amici I telescope from the early 19th century, which was originally in the Amici Dome at Arcetri Astrophysical Observatory.

Important Info

Location: 5 Largo Enrico Fermi, Florence.

Hours: Monday–Friday, 9am–6pm; Saturday evenings by appointment. Daytime tours at 10am; night-time tours at 6.30pm and/or 9pm depending on the season.

Admission: Entrance is by suggested donation.

Website: *www.arcetri.inaf.it*

Florence Cathedral in Tuscany, near Arcetri.

Arecibo Observatory

PUERTO RICO

Arecibo Observatory in Puerto Rico is one of the most well-recognised telescopes in the world, and one of the largest as well. Arecibo was featured in popular culture throughout the 1990s, including the James Bond film *Goldeneye*, the movie *Contact* and *The X-Files* TV show. Since opening in the 1960s, Arecibo has focused on scientific research in the fields of radio and radar astronomy, including atmospheric science and the Search for Extraterrestrial Intelligence (SETI). The wide dish of the radio telescope at Arecibo is a main component of outreach to other worlds.

Arecibo is operated by the National Science Foundation in cooperative agreement with the University of Central Florida. This means that both professional astronomers and students can apply for usage time at Arecibo's radio telescope. Some of the major work at Arecibo has included accurately measuring the planet Mercury's 59-day orbit, work that confirmed part of Einstein's theory of general relativity (and earned the astronomers the Nobel Prize in Physics); generating a radar map of the surface of Venus; and discovering the first exoplanet, creating a long legacy of major astronomical discoveries.

In northwestern Puerto Rico outside the town of Arecibo, approximately 90 minutes by car from San Juan, the observatory is a relatively easy site to visit. The Science & Visitor Center, exhibition hall and auditorium educate visitors about past and present work at Arecibo. In addition, you can look out over the massive Arecibo telescope dish from an observation deck. On a VIP tour, a guide will take you to the edge of the dish to get a sense of its massive size first-hand, 1000 ft (305m) in diameter. The Caribbean Astronomical Society also holds astronomy night events one or two times per year at the Arecibo Science & Visitor Center, usually in Spanish.

Clockwise from top: Historic San Juan; a small cascade in El Yunque National Forest; Los Arcos Beach near Arecibo.

Be sure to experience some of the other natural wonders on Puerto Rico after a visit to Arecibo. El Yunque National Forest is managed by the US Forest Service and has hiking and climbing trails throughout a 44-sq-mile (114-sq-km) rainforest. If you'd rather sample the forest than walk through it, stop by El Portal Rain Forest Center, where you can traverse raised walkways that give you a view of the rainforest canopy.

Important info

Location: PR-Route 625, Arecibo.

Hours: Wednesday–Sunday, 10am–3pm.

Admission: Adults/seniors/children (5–12) $12/8/8.

Website: *http://naic.edu/ao/landing*

South African Astronomical Observatory

SOUTH AFRICA

From left: The SALT telescope at the South African Astronomical Observatory; the headland of the Cape of Good Hope.

You can find the primary observing facility in South Africa, the South African Astronomical Observatory (SAAO), in the Karoo region, approximately 230 miles (370km) inland from Cape Town. The semi-arid Karoo region is renowned for its clear skies, which are optimal for observing. Since SAAO was established here in the early 1970s, the facility has grown to house 15 telescopes that conduct both optical and infrared astronomy. Ideally situated for studying portions of the night sky that are not visible from other observatories, SAAO has

© Grobler du Preez / Alamy Stock Photo; © BlueOrangeStudio / Alamy Stock Photo

SAAO is headquartered in Cape Town on the grounds of the Royal Observatory at the Cape of Good Hope, which operated until 1971. SAAO hosts regular astronomy talks, lectures and events at its headquarters one or two times per month, a great option for visitors who are unable to make it up to the telescopes near Sutherland.

Important Info

Location: Hwy R356, Karoo, Namakwa.

Hours: Self-guided daytime tours, 9am–3pm, Saturdays; guided daytime tours, Monday–Saturday, 10.30am and 2.30pm; evening tours, Monday, Wednesday, Friday and Saturday.

Admission: Adults from R60-R100, children from R30-R50.

Website: *www.saao.ac.za, www.salt.ac.za*

several telescopes that operate as companions to similar ones in other parts of the world.

One of the most important telescopes at SAAO is the South African Large Telescope (SALT). This telescope has a hexagonal mirror roughly 36.5ft by 32ft (11.1m by 9.8m), making it the largest optical telescope in the southern hemisphere. The main specifications of the telescope are very similar to the Hobby-Eberly Telescope at McDonald Observatory in Texas but give a very different view of the night sky, with access to the Southern Cross and nearby star Alpha Centauri.

Unfortunately the research telescopes at the South African Astronomical Observatory are not open to the public at night. It's still possible to visit though, and tours are offered during the day and at night. Daytime self-guided and escorted tour options allow you to explore the visitor centre and have a guided tour of some research telescopes, including SALT; tours at night are focused on stargazing and astronomy, and you can have a look through two visitor telescopes.

If you're unable to visit SALT, it's still possible to get an understanding of what's happening at the facility. On the SALT website you can view live feeds, videos recorded daily and sped up to short clips of 20–30 seconds. The videos show the inside of the SALT dome from different angles, as well as views of the sky and the outside of the building. It's no match for a visit in person, however.

Teide Observatory

SPAIN

Visitors flock to Tenerife and the Canary Islands to soak up the sun. Less known is that the islands are also some of the best places in the world for observing and studying the sky, thanks to their geographical position and astronomical conditions, which are similar to those in Hawai'i and Chile. To take advantage of the site, the Teide Observatory was established in the early 1960s atop Teide volcano on Tenerife. Operated by the Instituto de Astrofísica de Canarias (IAC), it is home to nocturnal telescopes, radio telescopes and the largest solar telescope in the world.

To visit Teide Observatory, you'll need to pre-book a guided tour through the primary tour operator, Volcano Teide Experience; you can't visit the observatory without a guide. There are two options, a 90-minute tour that includes a guided walk through the observatory grounds and a solar observation session, and an 8½-hour tour that features solar observation, stargazing and a night-time observation session with a telescope. Tours are offered in both Spanish and English, but you need to book the correct day and time for your preferred language.

For amateur astronomers with specific astronomical theories they'd like to try to prove through observation, Teide Observatory is one of the few places where you can request access to use a professional telescope. Applications for the opportunity to work on the Mons 20in (50cm) reflecting telescope must be filed at least 30 days in advance, and you'll need to provide the basic information on your intended observations and prove your ability to correctly use the telescope. The dark skies above the volcano are also ideal for independent stargazing to see 83 of the 88 recognised constellations.

Clockwise from top: Teide Observatory at night; Teide Observatory below the Milky Way; Tenerife in the Canary Islands.

If you're visiting the Canary Islands, La Palma is another island destination for enthusiastic astrotourists. It is home to the Roque de los Muchachos Observatory, which became the site of the Isaac Newton Telescope when it was moved from the Royal Observatory in Greenwich in 1984. Over a dozen telescopes are located at Roque de los Muchachos Observatory, which was also an alternative site for the TMT.

Important Info

Location: San Cristóbal de La Laguna, Santa Cruz de Tenerife, Canary Islands.

Hours: Teide Observatory has no public open hours; tour hours are variable.

Admission: From €21 for shorter tours; from €56 for longer tours.

Website: *www.iac.es*

CERN

SWITZERLAND

This page & opposite: CERN's striking Globe of Science and Innovation; CERN is home of the Large Hadron Collider (LHC).

Where typical observatories concentrate on scanning the vast bodies of the heavens, the underground particle physics laboratory at CERN (Conseil Européen pour la Recherche Nucléaire, otherwise known as the European Council for Nuclear Research) is focused on the opposite end of the scale. Founded in 1954, this is the premier laboratory for research into particle physics. The results have huge implications for astronomers, as the researchers here study the basic nature of our universe. CERN is best known for the Large Hadron Collider (LHC), the largest particle accelerator in the world, a circular tunnel 16.7 miles (27km) long that is able to crash particles into each other at rates approaching the speed of light. Before the accelerator began operating, the great volumes of energy it would create caused some to worry that it might even generate a black hole! The LHC and the other accelerators and detectors at CERN are used to study how particles behave under extremes such as those believed to have existed at the beginning of our universe, the Big Bang. When researchers at CERN discover a new particle or try to prove theoretical ideas about how particles interact, they give us greater understanding of the universe, its formation and our place in it.

Parts of CERN are open to the public to visit and learn more about the work done at this massive research facility. These include permanent exhibits that teach visitors about particle physics and why researchers study particles, as well as exhibits that introduce you to the Large Hadron Collider and our understanding of everything from the Big Bang onwards. Of interest to computer science fans, CERN is also responsible for some major leaps in computing, including contributions to the creation of the World Wide Web. CERN offers free guided tours in both English and French throughout the week, as well as tours for student groups. Visitors can see the technical centre for the 7716-ton (7000-tonne) ATLAS detector, 328ft (100m) below the surface, where the existence of the previously theoretical Higgs boson particle was tentatively confirmed. Its discovery led to a Nobel Prize and gave confirmation to the Standard Model of particle physics, which outlines the four fundamental forces at work in the universe: the strong force, the weak force, the electromagnetic force and the gravitational force. These are the building blocks of everything from the stars to the galaxies.

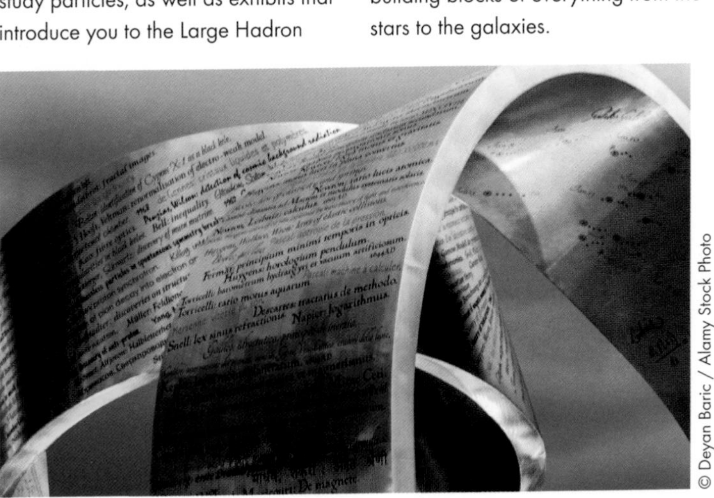

© Deyan Baric / Alamy Stock Photo

Inside the Large Hadron Collider it's colder than outer space. The interior of this massive loop is kept to a chilly 1.9°K (-456.3°F/-271.3°C). By contrast, space is a comparatively balmy 2.7°K (-454.8°F/-270.5°C). The temperature keeps the massive superconductor electromagnets cool as they work to accelerate proton beams.

Important Info

Location: Esplanade des Particules 1, Geneva.

Hours: Monday–Saturday, 8am–5pm; tours are offered Monday–Saturday, 11am–3pm.

Admission: Free, but reserve your tickets in advance.

Website: *https://home.cern, http://visit.cern*

Royal Observatory, Greenwich

UK

Rising like a beacon of time atop Greenwich Park, the Royal Observatory is home to the prime meridian (longitude 0° 0' 0"). Tickets include access to the Christopher Wren–designed Flamsteed House (named for the first Astronomer Royal) and the Meridian Courtyard, where you can stand with your feet straddling the eastern and western hemispheres. You can also see the Great Equatorial Telescope (1893) inside the onion-domed observatory and explore space and time in the Weller Astronomy Galleries. Though the main astronomical work once done at the Royal Observatory in Greenwich has since moved to less light-polluted skies, it's still a major piece of the history of science. The Royal Observatory was originally commissioned in 1675 for navigational purposes, and its early concern was to find an accurate means of marine timekeeping and ascertaining longitude. For this purpose it began issuing the Nautical Almanac in 1766, establishing the Royal Observatory as the baseline – hence the prime meridian line that runs through the building, and Greenwich Mean Time, still based on the time zone of the facility. Today the Royal Observatory in Greenwich is a museum and public education facility as well as being home to the Peter Harrison Planetarium.

During the day, visitors can attend a show at the planetarium on topics such as the night sky, neighbouring

From left: The Astronomer Royal, Mr Frank Dyson, taking an altazimuth reading; sunset at the Great Equatorial Telescope.

© Chronicle / Alamy Stock Photo; © Tony Watson / Alamy Stock Photo

planets and dark matter, or visit the Great Equatorial Telescope in its dome. The Royal Observatory also has exhibits on historic astronomy and hosts astronomy nights throughout the year for the public, including astronomy workshops for young people and talks and courses for both children and adults. An annual Astronomy Photographer of the Year exhibition showcases astrophotography.

Beginning in 2018, the Royal Observatory in Greenwich again became a home for observation with the installation of the Annie Maunder Astrographic Telescope (AMAT). As the AMAT is only open to the public on a limited basis, book tickets in advance; typically they sell out.

When the Royal Observatory was established in 1675, King Charles II named John Flamsteed as the first Astronomer Royal. Flamsteed worked to create both a 3000-star catalogue and a star atlas during his time at the Royal Observatory, and is credited as the first person to record Uranus (though he thought the planet was a star). He was succeeded in his post by Edmond Halley, who is most famously known for the comet that bears his name.

Important Info

Location: Blackheath Ave, London.

Hours: Daily 10am–5pm.

Admission: Free to visit the Great Equatorial Telescope; museum access adult/child £19/9; extra fees for special events, talks and planetarium shows.

Website: *www.rmg.co.uk/royal-observatory*

Mauna Kea

USA

Mauna Kea, a dormant volcano on the island of Hawai'i (nicknamed the Big Island), is considered one of the best spots for stargazing and astronomy in the world. High elevation combines with atmospheric conditions that are optimal for ground-based observation. At nearly 13,796ft (4205m), Mauna Kea is literally above some of the atmospheric conditions like humidity, weather and light pollution that normally interfere with observation of deep-space objects. Drawn by these favourable conditions, there are now 13 observatories around the summit, supported by 11 countries.

Mauna Kea was initially leased to the University of Hawai'i in 1968 for the development of an observatory, although recently, the proposed use of the mountaintop for construction of the Thirty Meter Telescope (TMT) renewed an ongoing debate over how to protect Mauna Kea's status as a sacred site for Hawaiians. Mauna Kea, meaning 'white mountain' in Hawaiian, is the umbilical cord, or ***piko***, of existence in Hawaiian culture. The astronomy reserve surrounding the summit is home to several goddesses, hundreds of sacred sites and the revered Lake Waiau.

Given its unique geographic position, it's not easy to reach Mauna Kea and the observatories. Once you arrive on the Big Island, you'll need to travel to the Mauna Kea Visitor Information Station (VIS) to check in on road conditions and observing opportunities, as well as to browse its excellent First Light Bookstore. Located on the Mauna Kea access road, the VIS is the most accessible place to go stargazing on Mauna Kea, with public astronomy events several nights each week. It underwent a renovation in 2019 that improved visitor facilities as well as enhancing and protecting the resources of the summit.

Beyond the VIS, only cars with 4WD are permitted to make the ascent to the observatories on the peaks. Additionally, Mauna Kea is only open from 30 minutes before sunrise until 30 minutes after sunset, and the telescopes are not open to the public at night, limiting stargazing opportunities from the summit. Some observatories, including the Subaru

© Alberto Ghizzi Panizza; © Joe Belanger / Alamy Stock Photo

From top: Sunset at Subaru Telescope; Keck Observatory above the clouds.

Telescope and Keck Observatory, offer day tours and viewings; others allow advance reservations for night-time tours for astronomy professionals and observatory groups. Another way to visit Mauna Kea is with a tour group. Operators are licensed to lead summit tours (typically eight hours long). Tours generally either focus on sunrise or sunset/stargazing experiences. If you visit, behave respectfully, keeping in mind the importance of Mauna Kea to traditional Hawaiian beliefs.

Astronomy and mythology about the stars have deep roots in Hawaiian culture. Ancient Polynesians who settled on Hawai'i used the stars to navigate across the Pacific Ocean, and it was believed that the demigod Maui snared the sun and used his magical fish hook to pull the islands up from the bottom of the sea. Though much of that astronomical knowledge was lost after European explorers arrived in Hawai'i, some native Hawaiians still study and practise Polynesian astronomy to navigate when sailing.

Important Info

Location: Mauna Kea Access Rd, Hilo, HI.

Hours: Monday–Saturday, 9am–10pm.

Admission: Free.

Website: *http://ifa.hawaii.edu/info/vis*

McDonald Observatory

USA

In the wide-open spaces of West Texas, the sky seems infinite, and stars stretch to the horizon in every direction. In the midst of the surrounding Davis Mountains, the University of Texas at Austin operates the McDonald Observatory and its range of telescopes and instruments, watching the skies and welcoming visitors by day and by night.

While most telescopes are measured in centimetres, McDonald Observatory has three permanent telescopes greater than 6.6ft (2m) in diameter: the 33ft (10m) Hobby-Eberly Telescope, the 8.9ft (2.7m) Harlan J Smith Telescope and the 6.9ft (2.1m) Otto Struve Telescope. Together these three make the observatory one of the most powerful in the region.

By day, visitors can explore the Frank N Bash Visitors Center, open daily, with exhibits about the work historically conducted at the McDonald Observatory. The centre is also the check-in point for daytime tours of the telescopes and night-time star parties.

Star parties at the McDonald Observatory are quite popular, and tickets are limited; reservations are encouraged. The parties occur up to three times weekly year-round, and you can pair a 90-minute twilight party with the two-hour stargazing session at the on-site telescope park. The programming at these events, led by observatory staff, varies based on the time of year and moon phase. If you are travelling specifically to attend events at the observatory, consider which astronomical events you most want to see and plan accordingly. Typically star parties will focus on the most prominent astronomical sights, ranging from planets to galaxy clusters to meteor showers.

If you're flying into the region, El Paso is the nearest major airport, with daily flights on many major airlines. It's a three-hour drive from El Paso to the nearest town, Fort Davis.

Clockwise from top: Star party attendees forming a line; McDonald Observatory under the stars; Mt Locke, Davis Mountains.

Extend your trip and head south toward the Mexican border in search of even darker skies than you'll find at McDonald Observatory. Big Bend National Park and nearby Big Bend Ranch State Park are both certified by the International Dark-Sky Association as Dark Sky Parks.

Important Info

Location: 3640 Dark Sky Dr, Fort Davis, TX.

Hours: Daily 10am–5.30pm year-round. Star parties are usually offered on Tuesday, Friday and Saturday each week; start times vary based on sunset.

Admission: Daytime tours, adult/child (6–12) $8/7; night events, adult/child (6–12) from $12/8.

Website: *https://mcdonaldobservatory.org*

NASA Jet Propulsion Laboratory

USA

Better known to Americans as the site of the annual Rose Parade, Pasadena also hides one of the world's most impressive space laboratories. The NASA Jet Propulsion Laboratory (JPL) in California may not be home to the rocket launches or astronaut training of other famous NASA facilities, but many of the most important robotics and exploration missions of the 21st century have been managed by teams based here. These missions include the first US satellite in space, Explorer I; Mars explorations such as the Curiosity rover and InSight lander; the Cassini mission to Saturn and the Juno mission to Jupiter; and dozens more.

The campus employs 6000 people, and among other things, it's where mission scientists and engineers on the Mars rovers direct the rover movements on the far-off red planet. At the start of each Martian sol, or day, rover controllers at JPL send instructions about which direction to move in and when to take scientific measurements, while trying to protect the instruments from running into disaster. A 20-minute communication lag between Earth and Mars means a difficulty level significantly higher than just running a remote-operated vehicle! Even after contact was lost with the Mars Opportunity rover after a major dust storm in June 2018, controllers at JPL sent a daily wake-up song to try to connect via 'active listening'. By then Opportunity had covered 26.2 miles (42km) since its landing in January 2004. In its travels, it confirmed the one-time existence of standing water on Mars and also made a strong argument for the possibilities of remote exploration: the rover outlasted its original projected mission length of 90 days by over 14 years. All of those movements on the distant red planet were directed out of the offices here in Pasadena, Earth's mission control for the remote operations on our solar system neighour.

Like most NASA facilities, JPL is open

© Stocktrek Images, Inc. / Alamy Stock Photo; © Sundry Photography / Shutterstock

to the public with limited access due to the sensitive nature of some projects and missions the staff is always working on. Free public tours are offered on alternating Mondays and Wednesdays but must be booked in advance. Tours include a presentation titled 'Journey to the Planets and Beyond', which focuses on JPL missions and accomplishments, plus a tour of the von Karman Visitor Center, the Space Flight Operations Facility and the Spacecraft Assembly Facility. A visit to JPL is an absolute must for space nerds who find themselves in Southern California, and pairs nicely with a visit to Vandenberg Air Base.

From top: The Mars Science Laboratory rover, Curiosity, during mobility testing; NASA JPL operations.

People can spend their whole careers working on a single mission at JPL. These missions are some of the most iconic and important in human exploration of the solar system via unmanned rovers and fly-bys: Cassini's survey of Saturn, the Voyager 1 and 2 missions to the gas giant planets, and the journey of New Horizons to Pluto and beyond compete for stature with Mars Curiosity and Insight. JPL's campus features photos from these missions, and on a visit you may cross paths with the scientists who made them possible!

Important info

Location: 4800 Oak Grove Dr, Pasadena, CA.

Hours: Public tours are offered specific Mondays and Wednesdays at 1pm.

Admission: Free, but reserve online far in advance.

Website: *www.jpl.nasa.gov*

Rocket City

USA

Visitors on a guided tour of the Davidson Center for Space Exploration, NASA's US Space & Rocket Center.

Located in northwestern Alabama, the city of Huntsville is also known as Rocket City, and it's not hard to see why. Huntsville has a long, prestigious legacy in the US space programme, as it is where Wernher von Braun led NASA in developing and testing the technology that eventually put the Apollo programme into space and on the moon. The city now hosts NASA's Marshall Space Flight Center as well as the US Space & Rocket Center, home to Space Camp. As you drive

© Dave G. Houser / Alamy Stock Photo

into town, you'll pass a huge replica Saturn V rocket standing sentry for the city. The journey through the Space & Rocket Center goes from the roots of America's space programme to the space-based science missions of today and tomorrow.

Huntsville is home to the famous Space Camp, a public outreach and education experience that is part of the US Space & Rocket Center. Visitors can enlist as astronauts for the week to learn more about the type of work and effort involved in being an astronaut. Space Camp welcomes astronaut-wannabes of all ages, from children ages 9 and up through adults, families and corporate groups. With gravity-defying simulators, rocket-building workshops and alumni who went on to become astronauts and engineers, this camp is no mere marshmallow roast! While attending Space Camp, you'll participate in simulated missions including launches, landings and spacewalks, plus learn about NASA's past and future missions. The US Space & Rocket Center also has a full-scale Saturn V rocket on display. Volunteer docents, usually former NASA/aerospace employees, called 'NASA Emeritus' to honour their contributions, can guide you through the massive museum that details the phases of space exploration and NASA's plans moving forward.

Travellers should also visit NASA's nearby Marshall Space Flight Center by bus as part of their trip. Tour stops include the historic test stand where engines were put through hot-fire testing to ensure they could handle the stresses of launch, and the Propulsion Research and Development Laboratory where NASA works to develop new rocket and engine systems for future missions. The local Von Braun Astronomical Society also operates a small observatory and planetarium at nearby Monte Sano State Park, open to the public most weekends for viewing.

Robin Soprano is Director of Simulations at Space Camp. 'Our goal is to tell NASA's story but also looking toward the future', she says about their programmes. 'We're trying to inspire the kids who will be the ones to go to Mars: what does it take to live independent of Earth?' Space Camp kids can learn how to fly NASA Orion spacecraft and simulate Mars colony habitat missions.

Important Info

Location: U.S. Space & Rocket Center, 1 Tranquility Base, Huntsville, AL.

Hours: Daily 9am–5pm.

Museum admission: Adult/child (5–12) $25/17

Space Camp: Child (9–14): 6 days, $999 ($1199 for teens up to 18)
Adult: 3 days, $549.

Website: *www.rocketcenter.com, www.spacecamp.com, www.nasa.gov/centers/marshall*

Meteor Showers

Meteor showers are a visible record of astronomic life in our solar system. Caused by comets and asteroids that leave trails of debris as they transverse Earth's orbital path, the detritus shed by these comets and asteroids on their journey flames out in glory when it meets Earth's atmosphere. Each year, Earth crosses those trails on the same schedule, to the delight of meteor watchers. As the planet passes through one of these fields of debris and a meteor shower occurs, it's a reminder that there are many objects in our solar system, moving around and dancing in sync without stepping on each other's toes. In fact, meteor showers occur on every celestial body as the planets, moons, asteroids and comets move in their celestial dance.

As Earth crosses the paths of cometary debris, it enters the atmosphere at a higher frequency than one-off objects entering the atmosphere from space. Watching these spectacular shows doesn't require anything more than the right timing: they're visible to the naked eye. When small objects hit Earth's atmosphere, we see these meteors light up as so-called shooting stars as they burn up on their fiery passage. While 'shooting star' may be a misnomer for the event, which is unrelated to any stellar phenomena, there's no denying the magic of seeing a bright object arc across the night sky.

Meteor showers are typically caused by comets, and less commonly by asteroids, but what's the difference between the two? Comets, which are responsible for the majority of meteor showers, are small solar system objects that are comprised primarily of ice and dust. They may also have an atmosphere, or tail, made of ice, dust and rocky particles. Comets in our solar system typically originate in the Kuiper belt, beyond Neptune. Asteroids, by contrast, are considered 'minor planets' in our solar system and have a rocky, mineral core. Typically they orbit the sun in a similar way to the other planets, and there is a large asteroid belt between Mars and Jupiter where most observed asteroids can be found. Some asteroids merely travel through our solar system on their journey, but these are less likely to leave a debris path notable enough to cause an annual shower.

Some meteor showers occur on a regular basis, including the majority of those in this section. There are also 'periodic' meteor showers, which occur less consistently: some years they may be very active, whereas in other years there may be no visible meteor behaviour at all (the Draconids are one such periodic meteor shower). Additionally, meteor showers are generally visible in parts, but not all, of the world, depending on the position of the debris path. It's their very unpredictability and brief lifespan that makes them so magical to view.

To see a meteor shower, plan to view in the hours between midnight and dawn. Most showers will experience the greatest visible activity at that time since the skies will be dark and most radiant points will be higher in the sky. But while there is a radiant point, identified as a constellation or collection of stars that meteors seem to appear from, don't look directly at it when trying to view meteor showers. You'll have a better chance of seeing meteors if you look in the space around the radiant point and, in some cases, throughout the whole night sky (as with the Leonids). It's optimal to try to view a meteor shower on its peak days, when the most meteor activity will occur; there are often forecasts available from NASA ahead of time that provide more detailed estimates. Increase your chances of seeing a 'shooting star' by planning ahead, but be aware that the dates may shift by one or two days on any given year. Always avoid nights with a full moon, or observe before the moon has risen or after it sets. Oh, and don't forget to make a wish!

© Belish / Shutterstock; © RGB Ventures / Alamy Stock Photo; Previous spread: © Haitong Yu / Getty Images

Quick Definitions

Comet: A small solar system object comprised primarily of ice, but which may also have dust and rocky particles in it

Asteroid: A small object comprised primarily of rock, usually from within our solar system but sometimes from beyond it

Meteoroid: A very small space object made of rock or dust, bigger than a molecule but smaller than ~330ft (100m) in diameter

Meteor: A meteoroid that has come into contact with Earth's atmosphere

Meteorite: A meteor that passes through Earth's atmosphere without burning up and makes impact on Earth

A composite image of the Perseids meteor shower taken in 2016.

Quadrantids

DECEMBER 28–JANUARY 12

While many cultures celebrate the start of each Gregorian calendar year with fireworks, did you know there's a show of celestial pyrotechnics you can watch instead? Though less well known than some other meteor showers during the year, the Quadrantids occur during the start of the year and typically peak on January 3.

At their peak the Quadrantids are as active a meteor shower as others, including the Perseids in August and Geminids in December. Most people don't catch this meteor shower, though. This is in part because the peak of activity is much shorter than in other active meteor showers, typically lasting just eight hours and sometimes occurring during the middle of the day for certain time zones. During this peak window, you may see as many as 50–100 meteors per hour. In general you can expect to see activity

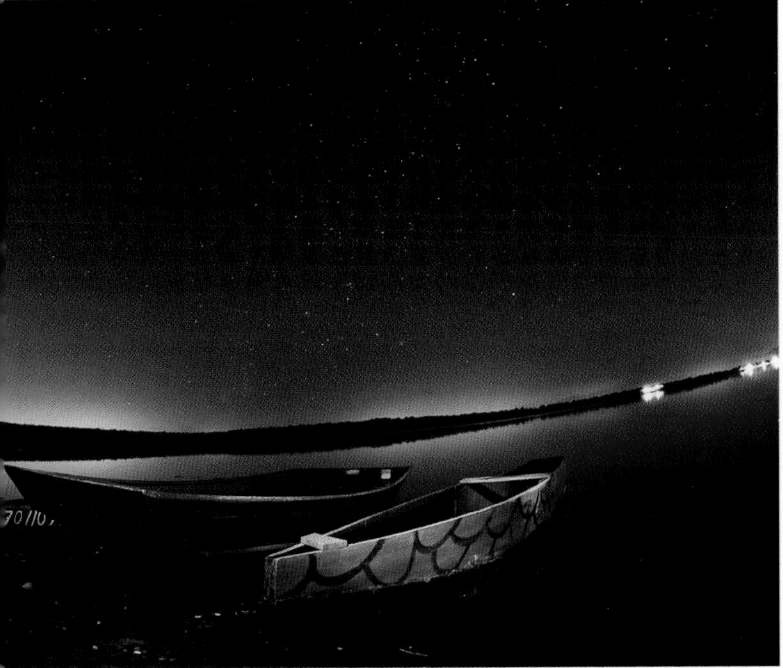

in the range of 10–15 meteors per hour on other nights of the shower.

The source of the Quadrantid meteor shower is not completely known, though the meteor shower has been observed for centuries. It is hypothesised that the Quadrantids are a relatively young meteor shower, beginning within the last 500 years, and either related to a comet (now called C/1490 Y1) originally observed by Chinese, Japanese and Korean astronomers or to the asteroid 2003 EH1. Adding to the mystery around this meteor shower, the constellation this shower is named for is now obsolete. The name Quadrantid comes from the constellation Quadrans Muralis, created in the late 18th century

From left: The Big Dipper, or the Plough; waiting for the Quadrantids in Miacatlán, Mexico.

as a depiction of a quadrant used in astronomical mapping but was absorbed into the constellation Boötes (the Ploughman) in the early 20th century.

To find the Quadrantids, look for the Big Dipper (also called the Plough). Following the 'handle' of the constellation, you can see the origin point for most meteors in the space between the final star and the constellation Draco. Another way to spot the Quadrantids is by looking for the orange giant Arcturus, the fourth-brightest star in the night sky, near which these meteors appear to radiate.

Although the Quadrantids meteor shower was not observed from the now-defunct constellation Quadrans Muralis until 1825, this constellation was named by French astronomer Jérôme Lalande in 1795. He named the constellation after an astronomical tool: the quadrant. Early astronomers used this tool to observe and plot star positions, giving the Quadrantids a deep astronomical heritage even though their object of origin is not fully understood.

Important Info

Dates: December 28–January 12

Typical peak dates: January 3–4

Most visible: Northern hemisphere, but partially visible as far as S 50°

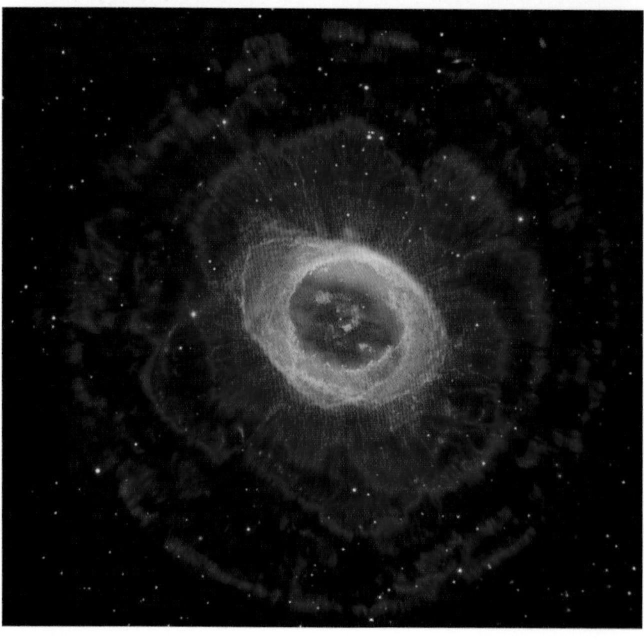

Lyrids

APRIL 16–26

The Lyrids in mid- and late April each year are the meteor shower with the longest recorded history, dating back to 687 BCE when they were noted by Chinese astronomers. While the Lyrids may not be as active as other meteor showers, peaking at an average of 15–20 meteors per hour on April 22–23, they make up for it in other ways. In a phenomenon known as a 'Lyrid fireball', some meteors are bright enough to cast shadows and leave trails of smoky debris in the sky for several minutes. The Lyrids can also have an increase in intensity when the Lyrid dust cloud is impacted by other planets, though these unpredictable occurrences usually happen only once every 20 years. In impacted years the frequency of meteors can increase by anywhere between 5 to 35 times their normal rate; in 1802, 700 meteors per hour were reported!

The Lyrids originate from the orbital path of long-period comet C/1861 G1 Thatcher, next expected to return in 2276. Long-period comets may have an orbit that lasts anywhere between 200 and infinite years (cases where we have never recorded the return of the comet, or its trajectory means it will never return). Comet Thatcher has an orbital period around the sun of 415 years, and this relatively quick orbit creates one of the strongest meteor showers for a comet of its type.

To view the Lyrids, look around the meteors' visual point of origin near the constellation Lyra (the Lyre or the Harp). The star Vega, also known as Alpha Lyrae, is the fifth-brightest in the night sky and the brightest within Lyra; use it as a guide to find Lyra and then monitor that sector to see if a Lyrid fireball appears nearby. Too close to the radiant point, and meteors will seem foreshortened.While you wait, also close to Vega is the Ring Nebula, a favourite of amateur observers.

Clockwise from top: Lyrids meteor shower in the Sierra Nevada; a Lyrids meteor against the sky; the Ring Nebula glows red.

While a telescope won't help you see meteors, you might want to bring one out while watching the Lyrid meteor shower. The Ring Nebula is the best-known planetary nebula visible in the night sky, and it is located within the constellation Lyra. You don't need a professional telescope to view the Ring Nebula; a personal telescope at least 3in (7.6cm) in diameter can help you spot its distinctive ring shape.

Important Info

Dates: April 16–26

Typical peak dates: April 22–23

Most visible: Northern hemisphere

Eta Aquarids

APRIL 19–MAY 28

Before the Lyrid meteor shower has even ended in April, another begins, this one more visible in the southern hemisphere. The Eta Aquarids begin in mid-April each year and continue for several weeks into May. This meteor shower was first observed in 1870, though its object of origin, Halley's Comet, may have been observed as early as 467 BCE. During the Eta Aquarids' height you can expect to see 10–20 meteors per hour.

The Eta Aquarids are unique in that there is no sharp peak in meteor activity during the shower. Instead you can view up to one week of moderately increased activity around the 'peak' date of May 5–6. This gives observers the opportunity to see the Eta Aquarids almost every year, depending on the moon phases; a full moon during this week will reduce the chance of seeing meteors. As with the Lyrids, there are sometimes increases in meteor activity based on the movement of other planets in our solar system, but the Eta Aquarids are among the most consistent in activity and duration from year to year.

Based on the time of year and the radiant point of the Eta Aquarids, you need to be near the equator for the best viewing opportunity. Look for meteors that appear to originate near the centre of the constellation Aquarius, which will be in the southern sky for viewers in the northern hemisphere, and higher in the sky for viewers in the southern hemisphere. From the northern hemisphere, the Eta Aquarids may even appear to be 'earthgrazers', moving parallel and close to the horizon.

Clockwise from top: Eta Aquarids fireball and twin meteors over Half Dome in Yosemite National Park; two images of Halley's Comet.

Halley's Comet no longer orbits close enough to Earth to leave a trail of debris in the planet's path. This means that the Eta Aquarids will eventually stop occurring once the existing trails of debris have all been dispersed in our solar system. But don't count on that happening anytime soon. It may be millennia before there is no more dust from Halley's Comet left in Earth's orbit.

Important Info

Dates: April 19–May 28

Typical Peak Dates: Week surrounding May 5–6

Most Visible: Equatorial down to S 30°

Delta Aquariids

JULY 12–AUGUST 23

The Southern Delta Aquariids and the companion Northern Delta Aquariids are two of the newer, less-understood meteor showers that occur each year. First observed independently of each other in 1870 and 1938, the two meteor showers were not

© Alvis Upitis / Getty Images

associated until 1952. The Southern Delta Aquariids are considered the stronger of the two showers and the one that's easier to see.

The Southern Delta Aquariids can be seen at a rate of 15–20 per hour during the shower, peaking at around 18 per hour on July 28–29. While this rate is not as high as that of other showers, it's still considered a strong rate of meteors. Note that if you are attempting to view the Southern Delta Aquariids in the northern hemisphere, your rate of viewing will be much lower as some meteors will not be visible beyond the horizon.

The object (or objects) of origin for the Delta Aquariids has not been confirmed. The strongest hypothesis is that the sibling meteor showers are caused by the comet 96P/Machholz, which was first observed in 1986. Comet Machholz is considered one of the 'sungrazer' comets; it has a short orbit of just over five years and travels within a few thousand kilometres of the sun on each passing. It's possible that the comet is a remnant of a larger sungrazer that broke up, creating the dust pathway that causes the Southern and Northern Delta Aquariid showers.

To see the Southern Delta Aquariid meteor shower, look for the constellation Aquarius (the same place you searched for the Eta Aquarids). The radiant point of the Delta Aquariids is the star Delta Aquarii (also known as Skat), a relatively faint blue dwarf star near the 'centre' of the constellation. You'll need to be in a dark-sky place to spot Delta Aquarii, and the closer you are to the equator and southern hemisphere, the better chance you'll have of seeing these meteors.

The Northern Delta Aquariid meteor shower occurs later than its southern sibling. Look for this shower each August, with a peak around the same time as the Perseid meteor shower. If you spot a meteor during the Perseids that appears from a radiant point in the southern sky (rather than the northern sky, where the Perseids' radiant point is located), it's likely one of the Northern Delta Aquariids!

Important Info

Dates: July 12–August 23

Typical Peak Dates: July 28–29

Most Visible: Southern hemisphere

A Southern Delta Aquariids meteor against the Milky Way.

Perseids

JULY 17–AUGUST 24

The Perseids are perhaps the most consistent, documented and popular meteor shower that occurs each year. Peaking in the height of summer for the northern hemisphere, between mid-July and mid-August, this is the northern hemisphere's most accessible meteor shower of the year. It forms many people's introduction to meteor viewing and is a centrepiece of many young stargazers' memories.

The debris that forms the Perseid shower is left by the comet Swift-Tuttle, which makes a 133-year orbit through the solar system; the comet was last seen in 1992. The Perseids have been recorded dating back to 36 CE when the Chinese first observed them, but the connection between the meteors and Swift-Tuttle wasn't fully understood until the comet made its fly-by in the mid-19th century. Since that discovery, scientists have come to understand and reliably predict meteor showers,

© Lee Thomas / Alamy Stock Photo; © Pluto / Alamy Stock Photo

including the night on which the meteors will peak and the frequency of meteors you can expect to see. Most years the Perseids peak at a rate of 60–70 meteors per hour; during a 'meteor storm' or 'outburst year', they peak at anywhere between 100 and 200 meteors per hour.

Typically the Perseids peak between August 11 and August 13, but the meteor shower officially begins around July 17 and continues through August 24. Meteors in the Perseid shower appear to emanate from the constellation Perseus, easily visible in the northern hemisphere in summer. However, on peak nights it's common to see meteors throughout the sky.

Due to the usually favourable weather during the month of August in the northern hemisphere, you don't need special equipment or gear to view the Perseids (as you might for meteor showers in winter). Instead, check with your local astronomy club for events or head to a dark-sky location for an informal viewing party. Most astronomy groups and observatories will be able to advise you about local events; the Perseids are popular enough that many people make specific trips to view them. In an annual calendar with ample meteor showers, the Perseids are the star.

From left: The Perseid meteor shower passes through the night sky above Broadway Tower, Worcestershire; Perseid meteor shower time-lapse.

As this is the most bombastic meteor shower of the year, astronomers throughout history have learned about meteor showers by studying the Perseids. In 1866 Italian astronomer Giovanni Schiaparelli realised that the periodic comet Swift-Tuttle's debris trail was the source of the Perseids, becoming the first to identify the link between comets and meteor showers.

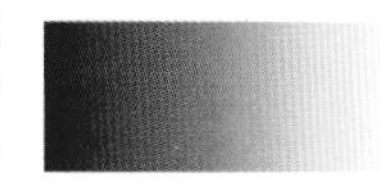

Important Info

Dates: July 17–August 24

Typical Peak Dates: August 11–13

Most Visible: Throughout the northern hemisphere

Draconids

OCTOBER 6–10

Among meteor showers, few are so impacted by the phase of the moon as the Draconids. This short shower occurs October 6–10 of each year, with a peak on October 8–9. Thus, if the moon phase is too bright (anywhere between the first quarter and last quarter) and depending on the time of moonrise, it may not be possible to see the Draconids well at all. On years when the moon is favourable, the Draconids are a short but exciting meteor shower.

The Draconids are caused by the comet 21P/Giacobini-Zinner, which is why they are sometimes also called the Giacobinids. Giacobini-Zinner has a six-year orbital period, and the trail of dust it leaves behind has caused a wide amount of variability in the Draconid meteor shower. Some years it may be possible to view up to thousands of meteors per hour; other years there may be hardly any visible meteors. Astronomers have attempted to predict the level of activity for the Draconid meteor shower each year, but so far they have not come up with a reliable model.

In an active year you may not need guidance to spot the Draconids: there may be so many meteors that it's impossible to miss them!

If you do need guidance, look for the constellation Draco, or the Dragon. Draco is a northern hemisphere constellation that is visible year-round. The tail of the Dragon can be seen between the Plough (the Big Dipper), part of Ursa Major, and Ursa Minor (the Little Dipper). The meteors appear to radiate from the Dragon's 'head', leading some astronomers to joke that Draco breathes fire during particularly active years.

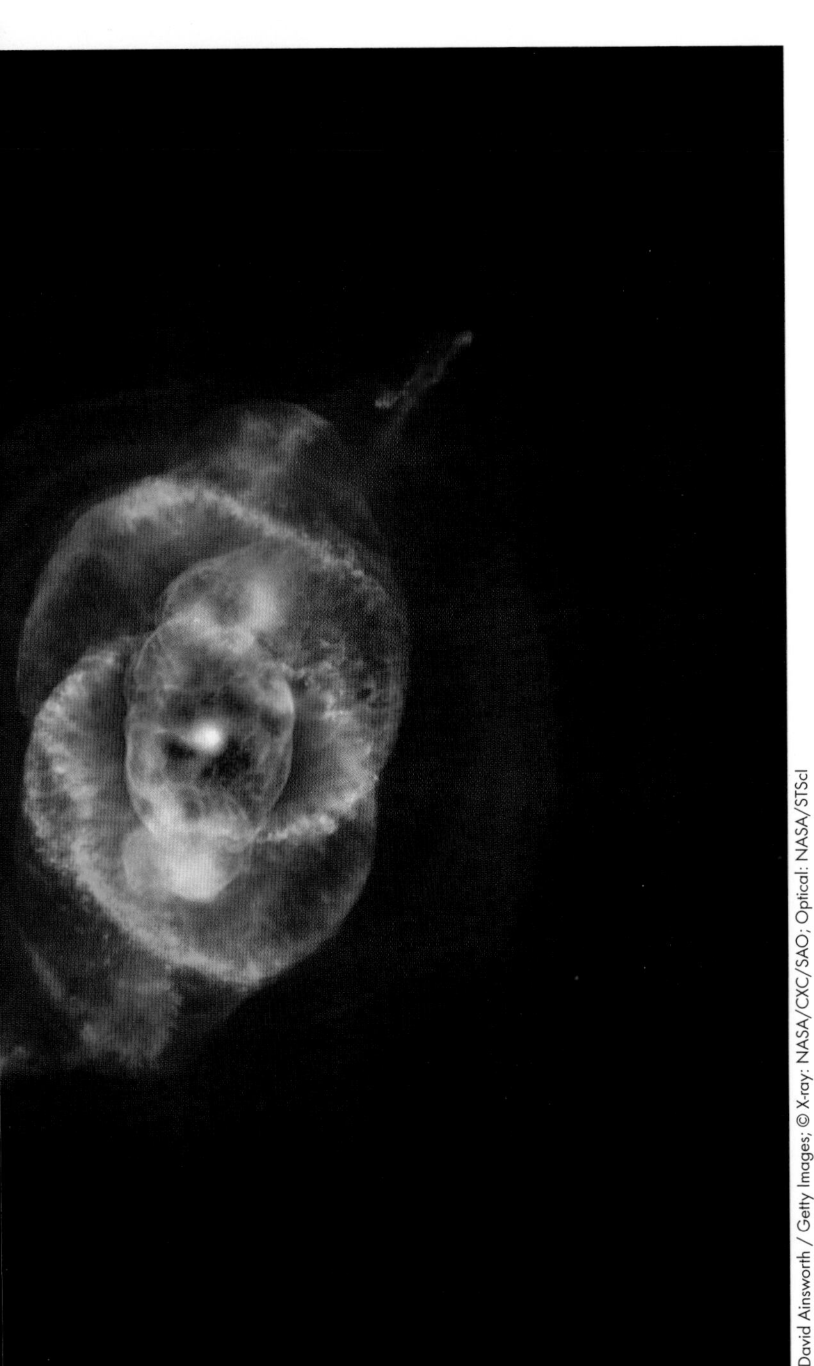

© David Ainsworth / Getty Images; © X-ray: NASA/CXC/SAO; Optical: NASA/STScI

You can view the Cat's Eye Nebula in the Draco constellation even during a year of few Draconid meteors. If you have a personal telescope, the Cat's Eye Nebula can be seen as a blue-green haze in the curve of the Dragon's back.

Important Info

Dates: October 6–10

Typical Peak Dates: October 8–9

Most Visible: Northern hemisphere

From left: A too-full moon obscures meteors; made up of eleven rings, the Cat's Eye Nebula is highly complex.

Orionids

OCTOBER 2–NOVEMBER 7

After the Perseids, the Orionids and the Geminids are the most popular and prolific meteor showers each year. The Orionids are especially great for viewing because they're easy to spot, appearing to radiate from the constellation Orion, visible worldwide due to its position on the celestial equator. The meteors in the Orionids shower are caused by debris left by Halley's Comet (as were the Eta Aquarids) on its 76-year orbit through our solar system. But unlike the Eta Aquarids, the Orionids are a more concentrated and active meteor shower with specific peak days that are optimal for viewing.

The Orionids were originally observed in the early 19th century and were attributed to Halley's Comet shortly thereafter. The meteor shower occurs each year between October 2 and November 7, and typically peaks around October 21–22. For viewers in the northern hemisphere, this means it's dark enough to enjoy the Orionids without the cold winter weather of meteor showers later in the year. In the southern hemisphere, the seasonal shift towards summer means it's possible to enjoy the shower without significant summer daylight interfering. During the peak, you can see up to 20–30 meteors per hour; in years with higher activity, there may be as many as 50–70 meteors per hour.

To view the Orionids, look for Orion in the northern sky. Orion is easily spotted by his 'belt' – the stars Alnilam, Alnitak and Mintaka – and the bright stars of Betelgeuse at his shoulder and Rigel at his foot. Meteors will appear to radiate from just above Orion's arm in all directions.

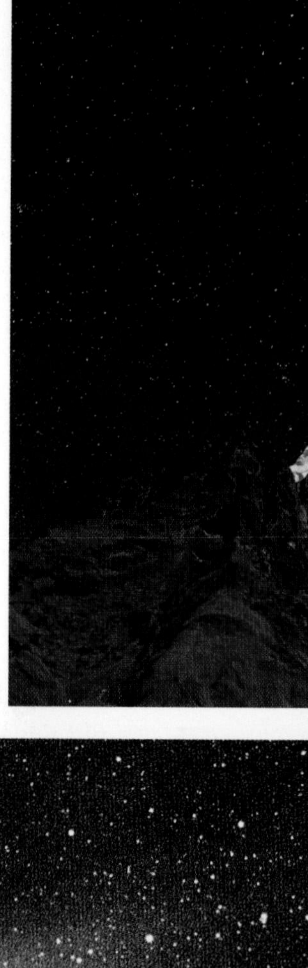

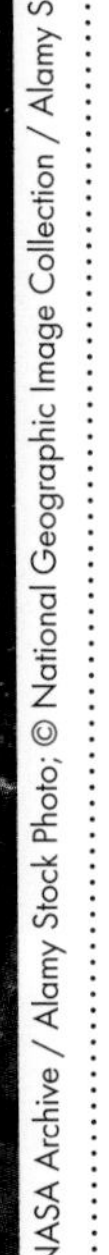

How can one comet cause two meteor showers? Based on the particular trajectory of Halley's Comet in our solar system, the inbound and outbound paths of the comet create two streams of debris almost exactly halfway apart on Earth's orbital path. When the Earth crosses through the outbound path, we see the Eta Aquarid meteor shower from mid-April into May; when it crosses the inbound path, we experience the Orionids in October.

From top: Orion rises above the Himalayas in the Sagarmatha National Park of Nepal; Halley's Comet, viewed in 1986.

Important Info

Dates: October 2–November 7

Typical Peak Dates: October 21–22

Most Visible: Worldwide

Taurids

SEPTEMBER 10–NOVEMBER 20

The Taurids are caused by a wide band of debris left over from a massive comet that is believed to have broken up between 20,000 and 30,000 years ago. This is the parent object for the comet and asteroid that separately give rise to the Taurids. One fragment, the comet Encke, is responsible for the Southern Taurid meteors; another, the eccentric asteroid 2004 TG10, gives rise to the Northern Taurids a month later. Thus, the Taurid meteor shower is comprised of the Southern Taurids first, followed by the Northern Taurids several weeks later. The Southern Taurids occur each year from September 10 to November 20 and are visible only in the southern hemisphere and equatorial regions, with meteor activity of the Southern Taurids typically peaking around October 10–11. During this time it's possible to see 5–10 meteors per hour as Earth passes through the trail of debris left by the comet each year.

As Earth moves into the densest part of the debris field from asteroid 2004 TG10, the Northern Taurids occur. This is typically between October 20 and December 10 each year, with the Northern Taurids' peak on November 12–13, though it's a less pronounced active period than what occurs for the Southern Taurids. During this time you can expect to see meteors at the rate of 5–10 per hour, radiating from a slightly lower part of the constellation Taurus.

While the showers have two different activity periods, they are part of the same broad band of debris. To view the Taurids in your hemisphere, look for Taurus, or the Bull, near the western horizon. It may help to think of the constellation Orion as pointing at Taurus with his bow and arrow.

Once you've spotted Taurus, the radiant points for both the Southern and Northern Taurids appear near the constellation. Keep your eyes peeled for fireballs, which were seen in activity bursts in 2005 and 2013; they may have a seven-year cycle. A Taurid meteor has even been observed impacting the moon!

From top: Taurus rises above the Himalayas; the Pleiades star cluster in Taurus, aka the Seven Sisters.

Before breaking apart some 20,000-30,000 years ago, the astronomical parent of Comet Encke and asteroid 2004 TG10 had a long history in our solar system, creating massive meteor showers that some have posited relate to Neolithic stone formations in parts of the world. Others have proposed that the split generated meteors so large that they caused a major extinction in the history of the Earth.

Important Info

Southern Taurid Dates: September 10–November 20

Southern Taurid Typical Peak Dates: October 10–11

Northern Taurid Dates: October 20–December 10

Northern Taurid Typical Peak Dates: November 12–13

Most Visible: Worldwide

Leonids

NOVEMBER 6–30

The flurry of meteoric activity in October continues apace with the Leonids, which begin while the Taurids are ending. Although the Leonids are not always the most active meteor shower, in certain years they manage to surpass the rate of all other meteor

Hubble snapped this view of M66, the largest member of the Leo Triplet. If you observe the Leonids with a telescope, you can sight this galaxy too.

© NASA, ESA and the Hubble Heritage (STScI/AURA)-ESA/HubbleCollaboration)

showers combined!

The Leonid meteor shower occurs when the Earth passes through the trail of debris left by the comet Tempel-Tuttle on its 33-year orbit through the solar system. During that time you can typically see between 15 and 50 meteors per hour; this number varies widely from year to year. In addition the Leonids occasionally produce a so-called 'meteor storm' in excess of 1000 meteors per hour. This occurs as part of comet Tempel-Tuttle's orbit and has been consistently measured since the mid-19th century. While the Leonids do not produce a meteor storm every year, when they do, it's historically notable.

There have been two significant Leonid meteor storms in recent history. The first occurred in 1833, when it was estimated that between 25,000 and 100,000 meteors were visible per hour on the peak night of the meteor shower. Newspaper articles about the event included reactions from such notable historical figures as Abraham Lincoln, Harriet Tubman, and Frederick Douglass. Another massive meteor storm occurred in 1966, helping researchers determine that these huge meteor events were likely related to comet Tempel-Tuttle.

The Leonids are visible in both the northern and southern hemispheres, appearing to radiate from the constellation Leo, the Lion. To spot Leo, look in the eastern sky for the bright star Regulus. The radiant point of the Leonids is a little higher in the sky, near the lion's mane. In the southern hemisphere you may want to stay up late or wake up for a morning meteor viewing session, due to increased summer daylight and the position of Leo in the sky.

Even in quiet years it's worth going out to look at the stars during the Leonids. With a pair of binoculars or an amateur telescope, you can spot the galaxies M65 and M66 in the constellation Leo. These two picturesque galaxies are visible under the Lion's body and are at a good angle for seeing the spiral shape of each.

Important Info

Dates: November 6–30

Typical Peak Dates: November 17–18

Most Visible: Worldwide

Geminids

DECEMBER 4–17

The last meteor shower of the year is one of the best. In fact the Geminid meteor shower, which occurs from December 4 to 17, is growing stronger and will likely outpace all other meteor showers in coming years. The Geminids typically peak December 14–15, and in an average year you can expect to see as many as 100–200 meteors per hour during that time!

© Benjamart Chiranurungsee / Shutterstock; © Stocktrek Images, Inc. / Alamy Stock Photo

From left: The Geminids meteor shower; capture of the Geminids streaking above China.

The rate of meteor activity during the Geminid meteor shower is increasing over time due to the originating rock comet's orbit (it will decrease again after 2200). As its intensity grows, it's possible the Geminids will become the most prolific meteor shower, popular for amateur and professional astronomers alike.

The Geminids are one of only two meteor showers caused due to an asteroid rather than a comet. In this case the Geminids arise from the debris trail of the asteroid 3200 Phaethon, often referred to as a rock comet (by comparison, most comets are icy). This asteroid has a 1.4-year orbit that takes it closer to the sun than any other asteroid in our solar system. Because 3200 Phaethon is in our solar neighbourhood, scientists study its transit to better understand the Geminids and other showers.

To view the Geminids, look for the constellation Gemini, the Twins. Gemini sits above Orion's shoulder in the night sky. Meteors will appear to radiate from the Twins' heads, but you'll have a better chance of seeing them if you look away from the radiant point. Geminid meteors are not as fast-moving as other meteor showers, making them easier to spot. The Geminids are also best viewed around 2am local time wherever you're trying to see them.

Under nice dark skies it's possible to spot an incredible astronomical object in Gemini with your eyes – no telescope or binoculars needed! M35 is an open cluster of several thousand stars near the foot of one of the Twins. Another way to spot M35 is by looking from the bright star Betelgeuse in Orion toward Gemini. M35 may look like a tiny cloud in the sky, but it's really composed of thousands of stars.

Important Info

Dates: December 4–17

Typical Peak Dates: December 14–15

Most Visible: Northern hemisphere, but also visible in the southern hemisphere

Ursids

DECEMBER 16–26

The calendar year ends as it begins, with a fiery show in the night sky. The Ursid meteor shower is relatively short, occurring from December 16 to 26 and peaking around the solstice, December 22–23. During the meteor shower you can expect to see 5–10 meteors per hour, depending on the year. Similar to the Quadrantids, the Ursids have a very short peak window, generally only 12 hours, in which the maximum meteor activity occurs.

The object of origin for the Ursids is Tuttle's Comet, first observed in the mid-19th century. Tuttle's Comet is unique among those that cause the major meteor showers because it is considered a contact binary: it is likely formed of two separate celestial bodies that were pulled together by gravity until they touched. It is the largest contact binary comet known, and the only one responsible for a meteor shower.

The Ursids are among the easiest meteor showers to spot, as they radiate from a point within Ursa Minor (the Little Bear) or often just from the Little Dipper. They appear to be directly above the Plough (also called the Big Dipper) in the sky, near the North Star. As the Ursids occur in the heart of winter for the northern hemisphere, look for meteors in the north sky, or directly overhead if you're far enough north.

From left: The Big and Little Dippers over Portal, Arizona; a hiker in Newfoundland gazes at the Big Dipper.

© Image Source / Alamy Stock Photo; © Alan Dyer VWPics / Alamy Stock Photo

The comet that causes the Ursids, Tuttle's Comet, is named after a notable American astronomer whose name you may recognise, Horace Parnell Tuttle. He also co-discovered comet Swift-Tuttle, responsible for the Perseids in August, and comet Tempel-Tuttle, responsible for the Leonids in November.

Important Info

Dates: December 16–26

Typical Peak Dates: December 22–23

Most Visible: Northern hemisphere

Aurora

There are endless reasons to be thankful for our sun. The gravity created by our star holds together the solar system, and it produces the light and heat we rely on for survival as a species. The sun also produces the spectacular light show we call the aurora.

Our sun is a dynamic star, powered by nuclear fusion, and particles from the sun are heated so rapidly that they escape the sun's gravity and move out in waves across the solar system, travelling at around a million miles per hour (400 km/s). This is known as solar wind, and its physics are not yet fully understood, though scientists continue to study our closest star to try to untangle the mystery. When these particles are especially intense from solar flares or coronal mass ejections, they collide with Earth's atmosphere to create the aurora, one of the most mesmerising natural phenomena on the planet. Seeing the aurora is one of the main highlights of winter tourism in the far northern and southern latitudes, and it's a once-in-a-lifetime experience for many travellers.

The aurora arise as charged protons and electrons hit atoms in the magnetosphere, or magnetic field, of a planet. Here on Earth, we call these the aurora borealis and aurora australis, depending on the hemisphere; other planets with a strong magnetosphere experience the aurora too, as the phenomenon has also been observed on Jupiter and Saturn. When these particles collide with atoms in our atmosphere, they emit light. Depending on which atoms they hit, and at what depth in our atmosphere, the light varies in intensity and colour from white to red, purple to green, even blue when charged nitrogen is hit. The lights seem to dance as the solar wind 'blows' across our atmosphere.

The aurora borealis is the term used to describe the aurora viewed in the northern hemisphere. This name, given by Galileo, combines the name of the Roman goddess of dawn and the Greek word for north wind. The aurora borealis, or northern lights, is visible in most countries on the northern part of the globe; popular destinations include Iceland, Norway and Canada. Specifically, the aurora borealis occurs around what is known as the auroral oval, which forms a ring around the North Pole approximately over the Arctic Circle, roughly overlapping with the same region that experiences the midnight sun in summer. Occasionally, strong magnetic storms mean destinations further south can experience the aurora, such as the northern parts of the United States and continental Europe.

The aurora australis, also called the southern lights, occurs in the southern hemisphere and is caused by the same phenomenon. A southern auroral oval occurs around the South Pole, and those destinations that fall underneath it are the ones most likely to experience the aurora australis. Visible primarily in Australia (especially Tasmania), New Zealand and Antarctica, the southern lights have occasionally been seen in the Patagonia region of South America and on islands off the South American coast, including the Falkland Islands and South Georgia and the South Sandwich Islands.

When it comes to viewing the aurora, there are several travel considerations to keep in mind. Above all you'll need to travel at the right time of year. While the aurora occurs in both hemispheres year-round, it is only visible when the sky is dark enough. This typically means that you'll need to view the aurora during the winter months of the hemisphere you're visiting for longer nights: November through February in the northern hemisphere, June through August in the southern hemisphere. Given that, you'll need to plan ahead and dress warmly for any aurora-viewing session. Most aurora destinations experience snowy, cold winter weather, and you should wear proper layers to enjoy the experience.

Finally, it's important to remember that as natural phenomena, the aurora are somewhat unpredictable. That's because solar activity itself is erratic and centred on the flipping of the sun's two poles in a process that alters its magnetic field. This roughly 11-year solar cycle sees an increase in sunspot activity between solar minimum and solar maximum, with the aurora intensifying around solar maximum, the last time back in 2014. Some experts think we are heading into a longer solar minimum cycle than usual, bringing about fewer sunspots and flares. If true, the range of latitudes at which the aurora is visible might decline. Scientists have improved predictions by observing and measuring activity around the sun's fluctuating magnetic field, but it can't be readily forecast. It may be the case that you plan to visit an aurora destination, but the conditions don't cooperate. Allow two or three nights minimum for any aurora trip to increase the odds of the right conditions for a successful viewing.

Common Aurora Borealis Destinations:

- Alaska (USA)
- Canada
- Finland
- Greenland
- Iceland
- Norway
- Russia
- Sweden

Common Aurora Australis Destinations:

- Australia, especially Tasmania
- New Zealand

Alaska, USA

SEPTEMBER–APRIL

The northernmost state in the USA, Alaska is ideally situated for viewing the northern lights, even as far south as its largest city, Anchorage. It's also the only part of the country that can regularly expect to view the aurora borealis. Due to the lack of large cities (Anchorage is home to fewer than 300,000 people), it's easy to escape what little light pollution there is and find dark skies on a clear night. Guided single- and multi-day aurora tours are also offered throughout the winter months to help you chase the aurora, if that is your preferred travel style.

Fairbanks

As the largest major city in the northern part of the state, Fairbanks is the primary base for aurora-viewing tours in Interior Alaska. You can drive or fly

© Mrs. Loh / Shutterstock; © DCrane / Shutterstock

From left: Denali National Park; the aurora borealis glowing green over a luxury camping tent in Alaska.

from Anchorage to Fairbanks throughout the winter, then explore the surrounding countryside by car or with a guide.

Chena Hot Springs Resort is a popular spot to visit in Fairbanks year-round. In winter the resort offers dog sledding, an ice museum and packages with access to the hot spring tubs that allow you to view the aurora while enjoying a warm soak – if the conditions are right!

Denali National Park

While there is reduced access to Denali National Park during the winter months and accommodation options are limited (you may need to stay in the nearby community of Healy), Denali is the most accessible of Alaska's large national parks. The lack of development guarantees dark nights and easy viewing of the aurora borealis against the mountain profile of Denali (formerly called Mt McKinley).

By day, popular activities include dog sledding, cross-country skiing and snowmobiling; these can be done independently or prearranged through tour providers in the region that specialise in winter experiences.

Anchorage

Anchorage is the state's largest city and has all the major amenities needed during winter travel. The Anchorage Museum is a good attraction for those interested in learning about Alaska's history and heritage. Visiting during early February means you can also experience the Fur Rendezvous festival and the start of the famous Iditarod sled-dog race.

To view the aurora borealis near Anchorage, head out of downtown to smaller communities like Eagle River or Girdwood, and then visit protected lands like the Chugach National Forest to escape light pollution. In less than an hour from Anchorage, it's possible to reach dark skies equivalent to those much further from city development.

From left: A hut in snowy Alaska; the northern lights at Ski Land near Fairbanks.

© Sunny Awazuhara / Getty Images; © Piriya Photography / Getty Images

Author Valerie Stimac grew up in Eagle River, Alaska, a 15-minute drive from Anchorage. Her top aurora-viewing locale is Beach Lake. There are several parking spots and all are protected from light pollution by trees. It's also a small-enough 'locals-only' spot where you might have the whole place to yourself on a night of aurora viewing.

Important Info

When to visit: Snow flies in Alaska from September through April. February is a great month to visit due to the famous Iditarod race.

Aurora alerts: The University of Alaska in Fairbanks offers an aurora forecast (*https://www.gi.alaska.edu/monitors/aurora-forecast?*).

Website: *www.anchorage.net*

Canada

OCTOBER–MARCH

Planning a trip to Canada to see the northern lights is less about attempting to visit the whole country and more about selecting the cities, provinces and territories you want to experience. Canada is a choose-your-own-adventure kind of destination, whether you sail the coastal waterways of British Columbia, drive the Icefields Parkway in Alberta, explore the big cities in Ontario and Québec, or enjoy fresh Atlantic seafood in Nova Scotia or Newfoundland. Even with several months or a whole winter season, it would be difficult to visit all of the places where you can see the northern lights in Canada, let alone see everything else this vast country has to offer. Instead, focus your itinerary by considering the daytime activities you'd like to experience in addition to aurora viewing at night.

Travel within Canada is relatively easy by car, making it a great destination for the independent traveller. Most highways are fully maintained through the winter months, so with 4WD, it's possible to plan and execute your aurora trip entirely on your own. There are also a large range of northern lights tours available from most cities and throughout all provinces and territories; these range from tours lasting a few hours to multi-day, multi-destination guided itineraries. Most tours will operate within a single province or territory, or travel within two, given Canada's large geographical area.

Wood Buffalo National Park, Alberta/Northwest Territories

Wood Buffalo National Park is the largest national park in Canada and a certified dark-sky preserve. Located in the far northeastern part of Alberta and the southern Northwest Territories, Wood Buffalo National Park is open year-round, with visitor centres in Fort Smith (Northwest Territories) and Fort Chipewyan (Alberta). The excellent Thebacha and Wood Buffalo Dark Sky Festival occurs every year in August. While this might seem too early in the year to see the aurora, the festival's days are filled with education and research programming, with stargazing and (maybe) aurora spotting at night.

Northern lights over Miles Canyon, Whitehorse.

© All Canada Photos / Alamy Stock Photo

Whitehorse, Yukon

Whitehorse is the capital of the Yukon territory and an ideal base for viewing the aurora. While it's difficult to see the northern lights from Whitehorse itself due to light pollution, it's possible from the surrounding region. Tour operators offer a variety of options from evening tours to multi-day excursions; Whitehorse is also one of the most popular cities in Canada from which to book an aurora tour. If you want to explore further, Kluane Lake is a two-hour drive north from Whitehorse and a great base for exploring the nearby national park

and glacial ice field. An even more extravagant and intense option is to book a seat on an Aurora 360 charter flight (aurora-360.ca) from the Yukon. Weather- and aurora-dependent, these flights get passengers closer to the northern lights than any land-based viewing point ever could. They fly in search of the aurora for a unique experience.

Churchill, Manitoba

From a base in Churchill on the shore of Hudson Bay, travellers can chase the aurora by night – and view polar bears by day. Thanks to the efforts of the Churchill Northern Studies Centre, which studies Arctic species and phenomena including polar bears and the northern lights, visitors can learn about the Arctic climate and

From left: A Yellowknife lake reflects the aurora borealis; a polar bear along the Hudson Bay coast.

© Jason Simpson / 500 px; © Robert Postma / Getty Images

. Churchill also sits under
oval, making it a prime
for viewing the northern
ever the skies are dark
d the weather is clear.

fe, Northwest Territories

of the Northwest
Yellowknife is another
n for a city in which to base yourself while exploring northern Canada in search of the aurora borealis. The city is on the northern shore of Great Slave Lake, and it's possible to strike out to the east or west. In both directions you'll find limited human development and low light pollution, making Yellowknife a good base for independent travellers searching for the northern lights.

To learn more about the aurora, visit Churchill Northern Studies Centre in Manitoba. In the winter months this facility conducts aurora and Arctic weather research and welcomes visitors interested in experiencing winter phenomena. You can view the aurora from under a transparent dome in the warm building. In other seasons you can see polar bears and belugas in the surrounding region.

Important Info

When to visit: To see the northern lights, visit between October and March. The best months will be those surrounding the winter solstice: December through February.

Aurora alerts: Aurora Watch ***(www.aurorawatch.ca)*** has a free email alert system to notify you if the northern lights are predicted or spotted in Canada.

Website: ***www.destinationcanada.com***

Finland

SEPTEMBER–APRIL

Of the Nordic countries, Finland is perhaps the most underrated northern lights destination, drawing fewer travellers despite its equally compelling aurora-viewing opportunities. Don't let that deter you: in Finland you can embrace Finnish sauna culture to warm up after a long session of viewing the northern lights, and there are many small communities in Finnish Lapland that welcome aurora borealis chasers. Many of these towns experience the aurora up to 200 nights per year, depending on solar activity and weather conditions.

While Finland is a less common destination on northern lights tour itineraries, some operators provide multi-day itineraries within the country. These generally allow you to try other Finnish experiences like dog sledding and enjoying the sauna during the day, and viewing the aurora by night. Finland also has comprehensive rail and highway systems that make it possible to travel independently throughout the country.

Rovaniemi

The capital of Finland's northernmost region, Rovaniemi, is a popular base for aurora tours. Dubbed the 'Official Hometown of Santa Claus', Rovaniemi is naturally a popular winter destination, and by day you can learn about the history of Santa Claus and enjoy winter sports or Lapland's wilderness. With Rovaniemi offering many amenities and a generous six month aurora window, weather conditions permitting, the city attracts the largest number of aurora-chasing travellers in the region.

Rovaniemi is accessible by car, train and plane from Helsinki and other European cities. It's a nine-hour drive from Helsinki or a 13-hour train ride.

Kemi

From Kemi, located on the shores of Bothnian Bay, you can look out across the water from Finland towards Sweden. Don't be surprised if you're distracted by the northern lights instead. With water to the south and relatively undeveloped Finnish Lapland to the north, Kemi enjoys limited light pollution, and it's possible to view the aurora borealis throughout the winter season. Kemi is also home to the

© Yoann JEZEQUEL Photography / Getty Images; © Roman Babakin / Shutterstock; Courtesy of Kakslauttanen

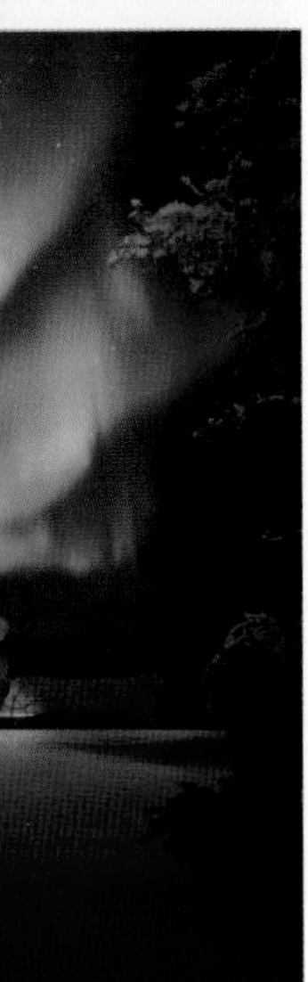

Clockwise from top: The aurora borealis in Lapland; reindeer in Rovaniemi; the glass igloos at Kakslauttanen Arctic Resort.

SnowCastle, similar to the Icehotel in Jukkasjärvi, Sweden, where you can stay in a seasonal ice structure each evening after viewing the aurora.

The town is more difficult to reach than other highlighted Finnish aurora destinations; it is most easily accessed by car. The final portion of the eight-hour drive from Helsinki winds along the coastline of Bothnian Bay, on a winding scenic road that was once a post road in the Middle Ages.

Kakslauttanen

The Kakslauttanen Arctic Resort (www.kakslauttanen.fi) has gained attention for its 'igloo' accommodations: glass-domed structures that allow travellers to literally sleep under the stars – or the aurora, if it is visible. This northern community is nestled between several national parks and other protected wilderness areas, and reduced development means less light pollution to interfere with your ability

© Wide Wings / 500px; © Danita Delimont / Getty Images

to see the northern lights. It's a major part of Finland's growth as an aurora destination.

Kakslauttanen is most easily reached by plane from major Finnish cities including Helsinki; it's an ideal base from which to experience winter sports by day and stargaze by night. A planetarium is in the works as well.

From top: Northern lights over Lapland; glass igloo for aurora viewing.

There are several Finnish legends around the aurora. The first, from the indigenous Sámi word for aurora, *guovssahasat*, suggests the aurora were spirits of the dead. In modern Finnish, the aurora are called *revontulet*, which has been translated as 'fox fires'. The mythical firefox causes the aurora with his tail, casting snow crystals into the sky and lighting the skies on fire. Other cultures in the region have similar myths in their past.

Important Info

When to visit: The months between November and February have the best viewing conditions.

Aurora alerts: Use the Auroras Now! service (*http://aurorasnow.fmi.fi*), run by the Finnish Meteorological Institute, ESA and the University of Oulu.

Website: *www.visitfinland.com*

Greenland

SEPTEMBER–APRIL

Greenland is among the least-visited countries from which you can see the aurora borealis; this is a shame as it's actually one of the best places for aurora viewing. Geographically part of North America, Greenland is heavily influenced by both European and Indigenous culture, and in the small communities where you can base yourself, you'll experience both.

Tour options are split along the western or eastern coasts: in West Greenland you'll travel between towns that offer modern amenities, whereas in East Greenland the most common option is a coastal cruise past the mostly uninhabited and protected part of the country.

If you're travelling independently, keep in mind that travel logistics in Greenland are a bit more complicated than in other countries, which in part explains the relative lack of aurora tourism to this massive, snowy country. If you're willing to plan ahead or book a group tour that arranges transport, you'll see a country considered to be among the world's top aurora destinations.

Nuuk

You might wonder if it's possible to see the northern lights from Nuuk, Greenland's capital city. With a population of only around 17,000 people, it has substantially less light pollution than you'd expect in a national capital. It also boasts the most cosmpolitan attractions in the country. While you can see the aurora best if you travel outside the city and escape what light pollution there is, it is entirely possible to view the aurora right in Nuuk.

If you plan to visit Nuuk, consider pairing it with time in Kangerlussuaq, Greenland's arctic aurora city; flight options are more flexible to and from Kangerlussuaq than they are to the capital city.

Kangerlussuaq

Base yourself in Kangerlussuaq if you're planning an aurora trip to Greenland. Kangerlussuaq is the most easily accessed community in

An iceberg in Greenland's North Star Bay.

Courtesy of NASA GSFC

Greenland, with flights to and from domestic as well as international destinations (Reykjavík, Iceland and Copenhagen, Denmark) operated both seasonally and year-round. Kangerlussuaq is located deep in a fjord in West Greenland, and is an ideal place to spot muskoxen. It is also the only place where you can access a road to the Greenland ice sheet, the massive ice formation that covers most of Greenland. Stay in Kangerlussuaq and book a tour out onto the ice sheet for some of the world's best aurora borealis viewing. Clambering over the ice is an unearthly experience.

From left: Greenland northern lights near Nuuk; a boardwalk leading from Ilulissat to the icefjord.

© Vadim_Nefedov / Getty Images; © Lottie Davies

the western coast of
like Kangerlussuaq,
'acts travellers to see the
Unesco World Heritage
'l as the aurora borealis.
graphers will find it a
' pleasant challenge to
photograph the aurora and massive icebergs floating in Disko Bay.

Some guided tours operate out of Ilulissat; others combine an itinerary in Ilulissat with nights in Kangerlussuaq. This may be in the running for the most remote spot to view the northern lights worldwide, which is saying something!

Many flights to Greenland include a layover in Iceland, so it's possible to plan an itinerary that allows you to see the aurora in both countries. Take advantage of the popular stopover option offered by Icelandair and you can visit Iceland for up to seven days as part of an itinerary to another destination.

Important Info

When to visit: Due to Greenland's high northern latitude, the aurora may be seen between September and April, a much longer season than in other destinations.

Aurora alerts: Aurora Service (***www.aurora-service.eu***) offers SMS/text message alerts for the aurora in Greenland.

Website: ***https://visitgreenland.com***

Iceland

SEPTEMBER–APRIL

© Ruth Nash / Shutterstock

Northern lights over Hella, Iceland.

Iceland has risen hugely in popularity as a travel destination, driven in part by the ease of viewing the aurora borealis throughout the country. There are guided tour options for those who want help hunting the aurora; many travellers opt instead to plan an independent itinerary and go around the country experiencing Icelandic culture by day and looking for the aurora at night.

The Golden Circle is a popular road trip itinerary looping 190 miles (306km) from Reykjavík and back; it can be done independently or with a guide. In winter many of the communities along the itinerary are well situated for viewing the northern lights. All three of the top destinations suggested in this section are on the Golden Circle. It is certainly possible to view the northern lights from other areas in Iceland, though. The southern part of Iceland is the most popular area for visitors, so if you're looking to escape the crowds, consider taking the Ring Road that circles the country and passes less-visited communities.

Thingvellir (Þingvellir) National Park

Located less than an hour from Reykjavík, Thingvellir National Park is the top aurora destination in Iceland. It is also a Unesco World Heritage Site that draws travellers year-round due to its close proximity to the Icelandic capital.

The national park, established in 1930, is home to stunning geography and has historic buildings and ruins to explore by day. At night, design-forward hotels in the area provide chic accommodations to keep you warm between aurora-viewing sessions, and increasingly cater to astrotourists.

Hella

Hella is one of the popular small towns outside Reykjavík where travellers flock to see the aurora. The local Hotel Rangá offers several amenities specifically for aurora chasers, including an aurora alert system, outdoor hot tubs to help you stay warm while viewing, and staff astronomers at their on-site observatory to answer any questions you have.

Hotel Rangá is sought-after enough to sell out each season, so plan ahead if you want to stay here; reserve several months in advance or be sure you book a guided tour that includes a stay at this property.

Skógar

Skógar is a destination that draws travellers year-round due to its proximity to Skógafoss waterfall, instantly recognisable as one of the top tourist destinations in Iceland. The 60ft (18m) waterfall is a popular spot for photographs.

The village is a base from which to see Iceland's stunning geography and enjoy the northern lights too. In the winter months the area is less crowded, and the surrounding cliffs are covered in snow, even as the waterfall pours into an icy lake below.

View of the northern lights with star trails at Skógafoss.

© Javen / Shutterstock

While there's much more to explore in Iceland than just the aurora, missing out on the chance to see them would be a lost opportunity. Thankfully, the Icelandic tourism industry is quite aware of just how much travellers want to view the phenomenon, with some hotels even offering an alarm service that wakes visitors up if the lights begin to show.

Important Info

When to visit: Visible in Iceland from September through April, the best time is the short nights of December through early February.

Aurora alerts: Aurora Service (*www.aurora-service.eu*) provides northern lights alerts via text message in Iceland. The Iceland Meteorological Office also offers aurora forecasts daily.

Website: *www.inspiredbyiceland.com*

Norway

OCTOBER–MARCH

For many years Norway has been the most popular destination for viewing the northern lights, aided by geography and the location of many small communities directly under the auroral oval. There are guided tour opportunities as well as resources for independent travellers to plan an aurora-viewing trip, and winter weather conditions make it likely you'll see the aurora as long as the solar activity is strong enough on your stay.

Svalbard

Svalbard is a group of islands set far north of the Norwegian coast, surrounded by the frigid Arctic Ocean, Greenland Sea and Barents Sea. Because of its isolated location well above the Arctic Circle, it takes additional planning to visit Svalbard, but it's also an ideal place to view the aurora from what feels like the top of the world. As Svalbard is so far north, the annual variance in daylight is dramatic: the city of Longyearbyen experiences 155 days with no sunlight each year! These days, from early October through early March, are the best time to try to see the aurora, especially the period surrounding the winter solstice in late December and early January.

Polar bears also call Svalbard home, and they exceed the population of humans on this archipelago. Brush up on polar bear safety before setting out on an aurora-viewing session.

Tromsø

Tromsø is one of Norway's most popular aurora destinations on the mainland, and you can view the northern lights right in the city on a good night. It also serves as a base for many northern lights tours in the region. By day, travellers can try dog sledding and snowshoeing, and visit a reindeer farm to learn more about life in the far north during winter.

The North Cape (Nordkapp)

The North Cape (Nordkapp) is the northernmost point of the European continent that can be reached by car. Though it's less

Aurora borealis in Tromsø over the Norwegian fjords.

common as an aurora destination than communities like Tromsø, Bodø and Senja, some aurora hunters visit for the bragging rights; they can say they've stood on this northern point and watched the northern lights dance in the sky. Road access is more difficult when there is heavy snow and wind in winter storms. There are several small communities in the Nordkapp municipality where you can base yourself for viewing the aurora, including Honningsvåg and Skarsvåg, the latter being the northernmost fishing village in the world.

Other aurora destinations in Norway include the Lofoten Islands, the Vesterålen archipelago, the communities of Bodø and Senja, and the Lyngenfjord region.

From left: Lighthouse on Gjesværstappan island in Nordkapp; the Midnight Sun at Nordkapp.

© alxpin / Getty Images; © Doug Pearson / Getty Images

In Old Norse myth during the time of the Vikings, the aurora represented a fire bridge to the sky built by the gods. One Norwegian chronicler explained the aurora as either an ocean surrounded by vast fires or the release of energy that turns glaciers fluorescent. Yet another ancient Norse legend held that the lights flickered out from the armour of warlike virgins riding horses across the sky. these shape-shifters of the night sky quickly take on fairytale qualities.

Important Info

When to visit: December to February is ideal; around the winter solstice, longer nights give more opportunity.

Aurora alerts: Aurora Service provides SMS aurora prediction alerts in Norway.

Website: *www.visitnorway.com*

Russia

NOVEMBER–MARCH

Russia is perhaps the least-visited aurora borealis destination in the northern hemisphere. Visa hurdles make it cumbersome for most travellers to visit Russia at all, much less to see the aurora. But for those willing to complete the necessary visa paperwork and fully plan an itinerary ahead of arrival, Russia is an unusual, uncrowded aurora destination.

Like Canada, Russia is such a large country that those travellers hunting the aurora here should narrow their focus and itinerary to a specific region in order to have the best chances of seeing the northern lights. While travel to Siberia sounds compelling, it's more difficult than most travellers are willing to undertake in order to see the northern lights, especially when there are more easily accessed destinations in western Russia.

Guided tours to see the aurora in Russia are typically longer than in other destinations; many combine sightseeing in major Russian cities with smaller towns or destinations where the aurora is better visible. Most independent travellers to Russia visit Moscow or St Petersburg, so where appropriate, consider adding travel time from these major cities to destinations where you can better see the aurora borealis.

Murmansk

Murmansk is generally considered one of the top destinations in Russia for seeing the aurora due to its high latitude and proximity to major cities like Moscow and St Petersburg. In fact most guided tour options will visit Murmansk as part of the itinerary. Located on a similar parallel to Tromsø, Norway, near Russia's border with Finland, Murmansk offers similar aurora-viewing conditions with far fewer crowds. By day you can explore local history and cultural museums to get a sense of Russian history. Murmansk is about a two-hour flight from Moscow or St Petersburg, making it an ideal destination for visitors. On the Kola Peninsula, it sits between the Barents Sea and the White Sea. In its forested depths is the Russian Samí town of Lovozero; this is one of the few communities to recover their indigenous traditions, which suffered greatly during the USSR years. Winter

Clockwise from top: The northern lights in Russia; traditional Salekhard dress; the distinctive Church of the Savior of Spilled Blood, St Petersburg.

finds Murmansk in near-permanent darkness because of its location within the Arctic Circle.

Naryan-Mar

Located far north near Russia's coastline on the Barents Sea, Naryan-Mar dates from the industrial era in Siberian history. A local monument in town remembers the first Russian city in the Arctic, Pustozersk, which is accessible only by snowmobile during the winter months. Naryan-Mar, home to roughly 20,000, is also a top destination to learn about the Indigenous Nenets culture and Russian history in the Arctic. It is a 2½-hour flight from Moscow to Naryan-Mar; daily flights are offered year-round on Utair. Visitors may find themselves sharing the flight with employees in the booming oil industry.

Salekhard

Further east in Siberia, Salekhard is another community great for seeing the aurora. Home to roughly 50,000 people, Salekhard is located on the Arctic Circle under the auroral oval, making it ideally placed to view the aurora. By day you can learn about indigenous history in the Arctic, and visitors can catch the annual Winter Show featuring an ice sculpture competition or learn about traditional reindeer herding in the region.

A view of the northern lights above Murmansk.

© ITAR-TASS News Agency / Alamy Stock Photo

In Slavic mythology the goddess Zorya represents the dusk and the dawn and is known as the Auroras (in some legends she is two goddesses). She guards the winged doomsday hound Simargl, who is chained to the North Star, Polaris, and keeps him from devouring the Little Bear constellation (Ursa Minor). Slavic legend holds that if Simargl ever breaks free and consumes the constellation, the universe will end.

Important Info

When to visit: Travel to Russia in the winter months between November and March if you want to see the aurora borealis.

Aurora alerts: Aurora Service (*www.aurora-service.eu*) offers SMS alerts for the aurora in Russia, particularly areas near Europe, including Murmansk.

Website:
www.russiatourism.ru/en

Sweden

NOVEMBER–FEBRUARY

Neighbouring Norway may receive more aurora-hunting travellers than Sweden, but don't let that fool you: Sweden – and similarly Finland – are both equally compelling destinations for seeing the northern lights. All three countries stretch northward along the Scandinavian Peninsula toward the auroral oval; in Sweden a major portion of the country is well-situated for viewing the aurora during the winter months.

As in its neighbours, there are a variety of aurora tour opportunities in Sweden; some cross into Norway or Finland, depending on the itinerary. It is also possible to travel independently while searching for the northern lights in Sweden.

Abisko National Park

The most popular aurora destination

© Johner Images / Getty Images; © Nick Upton / Alamy Stock Photo

From left: Abisko National Park; people entering the ice hotel, Jukkasjärvi.

in Sweden by far is Abisko National Park along the northern border between Sweden and Norway. While it's possible to view the northern lights throughout the national park on a clear night, Abisko is also home to Sweden's Aurora Sky Station. This mountaintop observatory is generally considered to be one of the best destinations in the world for viewing the aurora, due to elevation and a lack of light pollution.

If you're taking a guided tour through Sweden, it's likely Abisko National Park will be on the itinerary, and transport will be included. If you're planning to travel independently, consider taking the train from Stockholm. The 17-hour train ride will give you the chance to enjoy the countryside as you travel northward through Swedish Lapland.

Kiruna

Kiruna is one of the largest towns in northern Sweden, home to almost 20,000 people year-round. This small population helps make Kiruna an excellent base for aurora tours and viewing; it's easy to escape the city lights and get into the surrounding countryside to see the sky on a clear night, and many itineraries include a night in Kiruna for this reason. Easily accessible by the Swedish train system, Kiruna is also the site of the Esrange Space Center, a rocket range and research facility that studies the aurora borealis, among other astronomical phenomena. Where else can you see a rocket launch by day and the aurora by night?

Jukkasjärvi

Jukkasjärvi is a small community neighbouring Kiruna, made famous as home to the Icehotel, which was the world's first of its kind. This structure is built entirely from ice each year and draws travellers who want a full winter experience while travelling through Sweden. The Icehotel also offers aurora photography tours and 'safaris', where travellers board snowmobiles to head out into the Swedish countryside in the hope of viewing the northern lights.

Jokkmokk

The small Swedish Lapland community of Jokkmokk is the southernmost and perhaps least visited of the country's aurora destinations, located roughly along the Arctic Circle and ideally placed for viewing the aurora borealis overhead, though it's not as easily accessed as other destinations and is mainly reachable by car. By day Jokkmokk draws travellers interested in learning about the history and culture of the indigenous Samí people. During the winter months, the Winter Market, which has been held since the early 1600s, showcases a range of local Samí art and crafts.

The aurora borealis captured in a Swedish coniferous forest.

© Dave Moorhouse / Getty Images

Chad Blakely works for regional aurora tour operator Lights Over Lapland. 'The most amazing moments must be watching our guests see the northern lights dance for the first time', he said. 'I have seen grown men cry, young lovers embrace one another, impromptu wedding proposals, moments of uncontrollable laughter, and what can only be described as religious experiences. To be able to live vicariously through guests gives me an opportunity to relive my first aurora experience every night.'

Important Info

When to visit: November to February are ideal months.

Aurora alerts: Aurora Service provides aurora prediction alerts in Sweden by SMS.

Website: *https://visitsweden.com*

Additional Aurora Borealis Destinations

Besides the previously mentioned destinations, there are several other areas where it's possible to see the northern lights on a night when the aurora is particularly strong. Usually weather services issue alerts about these unusual occurrences so that locals can be sure to experience the event themselves.

Northern US States

On nights of strong solar activity, the aurora borealis has been seen in many of the northern states of the contiguous US. These include Washington, northern Idaho, Montana, North Dakota and South Dakota, Minnesota, Wisconsin, Michigan, and even parts of New England including Maine, upstate New York and northern Vermont.

While it is more difficult to predict when the northern lights will be visible in these states, it's possible (though unlikely) that winter travel throughout the northern US may give you the chance to see the aurora along the northern horizon on a cold, clear night.

Continental Europe

As in the northern US, countries in continental Europe can also experience the aurora borealis when the activity is strong enough and weather conditions are optimal. One of the best countries for this opportunity is Denmark, which stretches northward between the North and Baltic Seas, and especially the Faroe Islands.

Estonia, Latvia and Lithuania are also located far enough north on the globe that it's possible to view the aurora there. As with other countries mentioned, you'll need to venture out of major cities like Riga (Latvia) or Tallinn (Estonia) to reduce light pollution and increase your chances of seeing the northern lights.

The aurora borealis has also been seen at times in countries including the Netherlands, Germany, northern England, Ireland, Northern Ireland and Poland. In each of these countries, it's rare to see the aurora but on dark, clear nights with the right solar conditions, it's possible!

Aerial top view of the main meteorite crater in the village of Kaali, Estonia.

If travelling to Estonia, be sure to add Kaali crater to your itinerary. Perhaps proof of its powers of attraction, Estonia has one of the world's highest concentrations of documented meteor craters. At Kaali, 18km north of Kuressaare, is a 100m-wide, 22m-deep, curiously round lake formed by a meteorite at least 4000 years ago. There are a further eight collateral craters in the vicinity, ranging from 12m to 40m in diameter, formed from the impact of fragments of the same meteorite. This field of nine meteorite craters is one of the best examples of how small meteorites can change the landscape on impact. In Scandinavian mythology, the site was known as the sun's grave.

Scotland

Northern Scotland is another less likely but possible destination for viewing the aurora borealis. Transit options are limited in Northern Scotland or the Faroe Islands during winter, but it is possible to plan a trip that happens to coincide with a strong night of solar activity when you might see the aurora.

Australia

JUNE–AUGUST

Where better to see the southern lights, aurora australis, than in their namesake country? Australia is large enough to offer plenty of aurora-viewing destinations, though the island state of Tasmania is far and away the leader. If you visit Australia specifically to see the southern lights, plan ahead to drive away from the popular cities in search of dark skies where the aurora may be visible.

Although the southern lights don't follow the same seasonal variations as the northern lights, winter – between June and August when the nights are longest – is still the best season for viewing them. Snow isn't common along the coastlines of Australia and Tasmania, but it can get cold in the winter, so bring the appropriate layers to stay warm. Since the aurora australis isn't timed to a schedule, you may need to be out waiting for a long time before they appear.

© James Stone / Getty Images; © Kelvin Aitken /Alamy Stock Photo

From left: A star trail against the southern lights; a southern right whale, commonly seen in Bremer Bay.

Tasmania

Located off the southern coast of Australia, Tasmania is the best destination for seeing the aurora in the southern hemisphere. Even better, the southern lights are visible year-round with the right light conditions – though you may need to stay up quite late to see them in Tasmania's summer months of November to February.

Cockle Creek, on Tasmania's southern coast, is considered the optimal aurora-viewing destination; it's a two-hour drive south from Hobart, Tasmania's capital city. Nearer the capital, Mt Nelson and Mt Wellington are popular spots to escape the city lights and try to see the aurora. So are several of the south-facing beaches in the Hobart area, including Howrah Beach, Seven Mile Beach and Taroona Beach. As you look out over Betsey Island, the southern lights often arc over the sky. Tasmania's major airports are in Hobart and Launceston. Wild and isolated, the island also has a growing food and art scene, including the Museum of Old and New Art (MONA).

Victoria Coastline

On the Australian mainland, the scenic coastline of Victoria offers the best opportunities to see the southern lights on a good night. Base yourself in Melbourne but be prepared to drive a couple hours to escape the city lights. Locations including Point Lonsdale, Cape Schanck and Wilsons Promontory are isolated enough from light pollution that may otherwise interfere with viewing the aurora. If you're lucky, you can even catch the southern lights glowing over the majestic Twelve Apostles in Port Campbell National Park on the Great Ocean Road. Plan to stay in the area if you are chasing the aurora in one of these destinations, as you may spend several hours watching the aurora and not want to drive back to the city each night.

South & Western Australia

Along the southern coastlines in the states of South Australia and Western Australia, it is possible to see the aurora australis on a clear night of strong solar activity, though the odds are somewhat lower than in Tasmania or Victoria. South Australia and Western Australia are more sparsely populated than areas further east in Australia, which helps too. Gracious and relaxed, they offer a chance to bask under the southern sky away from the hustle and bustle (and light pollution) of the coastal cities to the east.

National parks and reserves punctuate the coastline. At some, including Cape Arid National Park and Fitzgerald River National Park, you can plan ahead to book a campsite for a night. If you want to expand your experience with nature, nearby Bremer Bay is a whale hotspot year-round, with orcas from January through March, and southern right whales in abundance from July to October; even humpbacks pass by.

The aurora australis from Thirteenth Beach, Barwon Heads.

© Terry Oakley / Alamy Stock Photo

While visiting Australia, be sure to plan a trip to Warrumbungle National Park in New South Wales (p40). This Dark Sky Park certified by the International Dark-Sky Association gets you well away from the city lights of Sydney, Canberra and Brisbane. Siding Spring Observatory is on the edge of Warrumbungle too. During the day, observe the eastern grey kangaroo population; at night, enjoy the starry skies.

Important Info

When to Visit: The winter months of June through August are ideal due to longer nights.

Aurora alerts: Aurora Service (*www.aurora-service.net*) offers a free SMS alert system when the aurora is expected or a sighting is reported in Australia or Tasmania.

Website: *www.australia.com*

New Zealand

JUNE–AUGUST

The aurora in the southern hemisphere may bear the name of aurora australis, but it's possible to see them outside of Australia too. New Zealand's South Island in particular is an increasingly popular destination for aurora chasers who want to sample the great winter sports opportunities in New Zealand by day, then see the southern lights at night. Skiers will find the best slopes outside Canterbury and Queenstown, and can travel down towards Invercargill for the best chance at seeing the aurora australis after burning up the powder.

New Zealand is slightly further south than Tasmania, aiding your chances of seeing the aurora. As in Australia and Tasmania, it's possible to see the southern lights here year-round, especially in the early morning hours when solar activity is particularly strong.

Invercargill

Invercargill is the southernmost city in New Zealand and home to 55,000 people. While viewing the aurora within Invercargill is difficult due to light pollution, it's an excellent base for seeing the aurora in the surrounding region.

From Invercargill, head east to The Catlins, a forest park with dark skies and good opportunities to see the southern lights on a clear night. Or head west towards Fiordland National Park; while the mountains may impede your view of the southern skies, the lack of development makes it a great dark-sky location for trying to view the aurora. You might also head south towards the coast and the community of Bluff. From here you can catch the ferry to Stewart Island or explore along the shoreline of the Foveaux Strait, with excellent views of the southern sky. Watch out for penguins as well. Visitors to the South Island may want to detour north for another scientific attraction too, this time of a geological nature: seeing the distinctive Moeraki boulders on Koekohe Beach create another kind of awe and wonder.

Stewart Island

The southernmost point in the country and a Dark Sky Sanctuary, Stewart Island is widely considered the best

Aurora australis at Meikel Johns Bay near Glenorchy.

© Gareth Davies / Alamy Stock Photo

place in New Zealand to see the southern lights. A one-hour ferry ride from Bluff on the South Island will move you far away from any light pollution on the mainland. Over 85% of Stewart Island is protected as part of Rakiura National Park, guaranteeing little human settlement and light pollution. On a good night of solar activity, the views of the aurora australis from Stewart Island are among the best in the world.

Aoraki Mackenzie Dark Sky Reserve

In 2012 the Aoraki Mackenzie Dark

Sky Reserve was established on New Zealand's South Island. This area, comprising some 1660 sq miles (4299 sq km), was established to protect the dark skies around Aoraki (Mt Cook) National Park and nearby Lake Tekapo, and to honour the importance of the night sky in Māori culture. Aoraki Mackenzie offers hiking and camping, as well as dark-sky tours and stargazing events. The reserve is a three-hour drive from Christchurch or a 3½-hour drive from Dunedin, rounding out your trip.

From left: Pier on Rakiura Track, Stewart Island; the Waipapa Point lighthouse against the aurora australis.

© slyellow / Shutterstock; © Arrested Light Photography / Alamy Stock Photo

If travelling throughout New Zealand as part of your aurora watching, consider visiting Aotea / Great Barrier Island (p80), located off the shores of the North Island near Auckland. This isolated island was certified as a Dark Sky Sanctuary in 2017.

Important Info

When to visit: You can see the aurora year-round, though the best opportunities will be in the winter months of June to August when the skies are relatively darker than in summer months.

Aurora alerts: Aurora Service offers text message alerts for aurora sightings in New Zealand.

Website: *www.newzealand.com*

Additional Aurora Australis Destinations

Because of the location of the continents and landmasses on Earth, there are fewer destinations from which to see the aurora australis than the aurora borealis. It's also more difficult to reach most southern aurora destinations during the winter season, when winter pack ice makes substantial parts of the southern oceans impossible to navigate.

It is possible to see the aurora in several destinations other than those previously mentioned, though much less common. Each of these locations will take additional planning to reach – but if you've got your mind set on viewing the aurora in every destination, these should be on your list.

Antarctica

As the major landmass at the southern end of the Earth, Antarctica is a natural contender for viewing the southern lights. Geographically, Antarctica is the best place to view the aurora australis, but it's also the most difficult place to visit in the winter months when the southern lights are most visible.

While scientific research and observation occur on Antarctica year-round, most options for travellers are limited to Antarctica's summer months of December through February. It is possible to find cruise and tour options in the shoulder seasons (November and March) when the aurora may be visible, but these options are less common and the chances of viewing the southern lights are less certain.

Patagonia

The Patagonia region in Argentina and Chile is far enough south that it's sometimes possible to see the aurora australis from here. However, it's rare that the southern lights are strong enough to be seen, and viewing them will require travelling into mountainous areas during the winter months. Many travellers base themselves in Ushuaia, Argentina, when attempting to see the aurora in Patagonia.

Falkland Islands

Located 300 miles (483km) off the eastern coast of Argentina, the

The aurora australis puts on a show for the Halley Research Station in Antarctica.

© Stuart Holroyd / Alamy Stock Photo

Falkland Islands are one of the most isolated places on Earth. Many visitors travel to the Falkland Islands to see penguins as part of an Antarctica itinerary; if you choose a shoulder-season cruise that visits the islands in April or August, you may see the aurora too. Flights are also offered intermittently from Río Gallegos in Argentina and Punta Arenas in Chile.

South Georgia and the South Sandwich Islands

Accessible only by boat and typically visited on Antarctica cruises, South Georgia and the South Sandwich Islands are even more isolated than the Falklands. Several cruise providers make port at the islands, though few travel in shoulder season when the aurora appear.

Eclipses

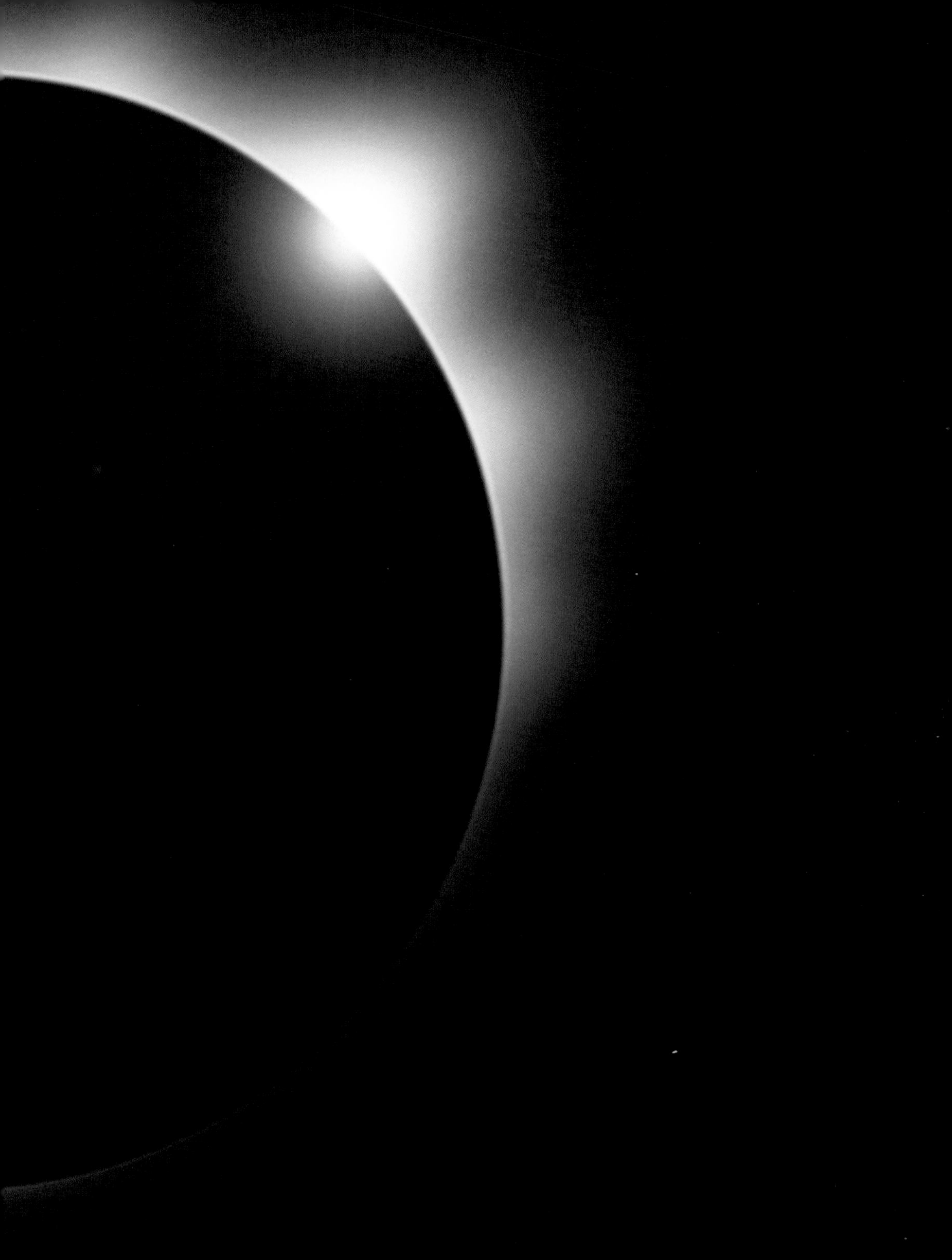

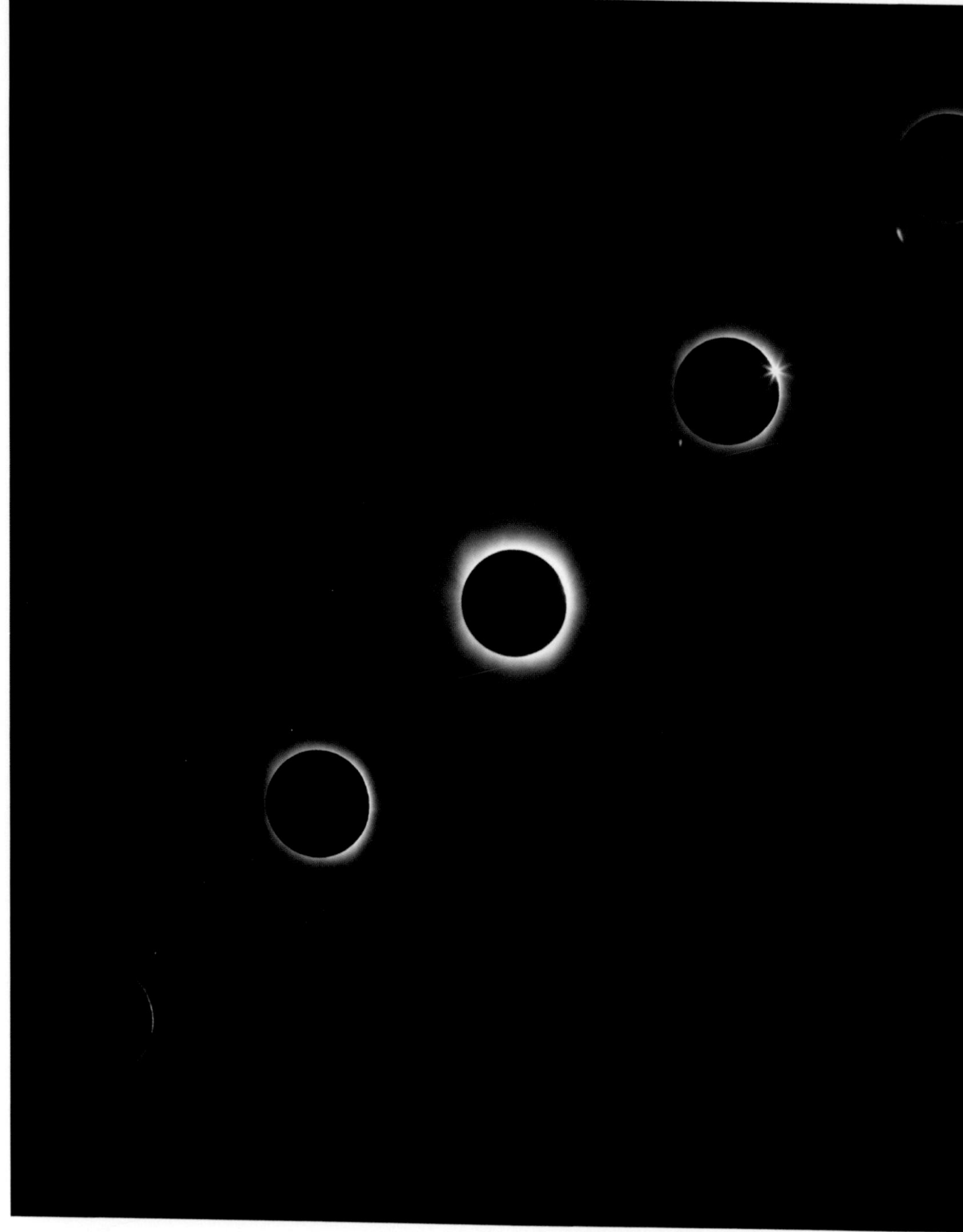

It's easy to take the sun for granted. Since well before the dawn of humankind, the sun has risen each morning, lighting the day, and set each evening to bring night. It controls the pace of the night hours for stargazing, viewing the aurora or trying to catch a meteor shower. By day we go about our lives under its constant glow. Through it all, the sun rises and sets on its consistent, endlessly perpetuated schedule. Until, that is, the disruption of an eclipse occurs.

As part of the celestial dance, Earth orbits the sun, and the moon orbits us. Eclipses occur when the shadow of the Earth crosses the moon, and vice versa. Broadly speaking, there are two types of eclipses: solar eclipses and lunar eclipses. A solar eclipse occurs when the moon passes between the sun and Earth, casting its shadow on a part of our planet. Depending on the distance between the moon and Earth at the time of a solar eclipse, the moon can actually cast up to two shadows: the penumbra, which is a wide, diffuse shadow, and the umbra, which is a smaller, darker shadow. Based on that distance and which of these shadows we experience, we have different names for solar eclipses: a partial solar eclipse, a total solar eclipse and an annular solar eclipse.

A partial solar eclipse occurs anytime the moon passes between the sun and Earth, creating a penumbra on part of the Earth. Technically every solar eclipse is a partial solar eclipse at some point.

A total solar eclipse occurs when the distance between the moon and Earth is such that the moon and sun appear to be the same relative size in the sky. During a total solar eclipse, the moon completely blocks the sun, creating both the umbra (which we call 'totality') and the penumbra (in which viewers experience a partial solar eclipse).

An annular solar eclipse occurs when the moon is relatively further from the Earth than during a total solar eclipse – the shadow of the moon does not completely block the sun. The common name for this eclipse is a 'ring of fire' eclipse, due to how its flare of light appears. No part of the Earth falls under the umbra, so there is no experience of totality. An annular solar eclipse is technically a special type of partial solar eclipse.

A lunar eclipse occurs when the shadow of the Earth falls on the moon. There are three types of lunar eclipses too: a total lunar eclipse, in which the Earth completely blocks the sun and our umbra covers the moon totally; a partial lunar eclipse, during which the Earth's umbra covers only part of the moon; and a penumbral lunar eclipse, in which the moon falls in the shadow of the Earth's less dramatic penumbra. In both a total and partial lunar eclipse, the moon in the Earth's shadow appears red, earning the nickname 'blood moon'.

Whatever you call them, eclipses have been of significance throughout human history. The earliest human record of an eclipse dates back to 2134 BCE, and there is evidence of them in some of the oldest preserved texts in the world. Often interpreted as an omen, their meaning depended on the culture and the circumstances during which the eclipse occurred. In China eclipses were associated with the health and longevity of an emperor, making it necessary to try to predict them; astronomers who failed to do so were often executed. The Babylonians would place a substitute king on the throne during eclipses so that the wrath of the gods would fall on him instead of the real king. In ancient Greece a solar eclipse marked the end of a war between two groups. Even today an eclipse is a significant moment for anyone who experiences it.

© Leo Patrizi / Getty Images

ECLIPSE-VIEWING SAFETY TIPS

While no special precautions are necessary to view a lunar eclipse, there are important safety considerations to keep in mind if you plan to travel and experience totality during one of the solar eclipses mentioned in this section, or if you plan to experience a partial or annular eclipse at a different time. In particular, at no point in a partial (or annular) eclipse can you look at the sun without proper solar eye protection. Even if the sun appears fully obscured by the moon, it's possible to damage your eyes if you view it without protection. During a total solar eclipse, look for the disappearance of what are called 'Baily's beads' or the 'diamond ring' to indicate totality has begun. Afterwards, and only during totality, is it safe to remove your eye protection and look directly at a total solar eclipse.

The ISO, or International Organization for Standardization, has specific recommendations for solar eclipse viewing glasses; look for glasses with the ISO 12312-2 rating on them. Be sure that any eye protection you purchase, including solar filters for your camera, is manufactured to this safety standard to avoid any risk to your eyesight.

With the right gear, you can safely watch the sky go dark as dusk descends in every direction, and the sun – ever-present in our daily lives – disappears behind the moon. First-timers may find it one of the most humbling and moving moments of their lives.

CATCH 'EM ALL

Almost every major region of the world will experience a total solar eclipse in the 2020s. Between four and six eclipses (solar and lunar, including partial, annular and total variants) occur every year based on our celestial dance with the sun and moon. Those elusive total solar eclipses will occur somewhere in the world roughly every 18 months. They will pass over some of the world's most remote regions – and some of the most populous. If you've never experienced totality, there will be several opportunities to do so during this decade.

While lunar and partial eclipses have their intrigue, you might want to prioritise the total solar eclipses listed here. While some are an adventure to get to in and of themselves, more central eclipse paths will be easy to plan for and well worth the advance effort; they're an experience like no other.

In December 2020, a total solar eclipse will pass over the southern portion of South America, mostly across Chile and Argentina. An eclipse in the same region occurred in July 2019.

In December 2021, penguins and extremely intrepid travellers can experience a total solar eclipse that will pass across Antarctica and parts of the southern Pacific and Atlantic Oceans. Because the eclipse will occur during the region's summer season, tour operators will likely take advantage of the chance to give travellers two bucket-list experiences: a trip to the Antarctic and totality.

In April 2023, a total solar eclipse will take place along a snake-like path through Oceania and Southeast Asia. Mostly passing over remote and rural parts of the region, this is another solar eclipse that will require travellers to make an investment of time and money in order to experience it. Major cities in the region will experience a partial solar eclipse too.

In April 2024, North America will experience its second total solar eclipse in less than a decade (the first was the 'Great American Eclipse' of August 2017). Travellers will flock from across the US and the globe to encounter totality on the eclipse's path through Mexico, Texas, the midwestern and Great Lakes states of the US, and northeastern Canada.

In August 2026, a total solar eclipse will take place in Europe as the moon's shadow passes over Greenland and Iceland and through central Spain. Much of the European continent will experience a dramatic partial eclipse too, guaranteeing a spike in astrotourism surrounding the event.

In August 2027, the regions of North Africa, the Middle East and far East Africa will see the second-longest solar eclipse of the century – a whopping six minutes, 23 seconds in length. While political tensions exist throughout parts of this region, the rarity of this event will likely encourage a tourism boom.

In December 2028, the final total solar eclipse of the decade will take place across Australia and New Zealand. If clear skies occur, citizens across both countries will experience a spectacular partial eclipse, and those in the cities of Sydney, Australia, and Dunedin, New Zealand, will experience totality.

South America

2020

Residents in southern Chile and Argentina are in for the treat of a lifetime: following the total solar eclipse in July 2019, the region will experience a second total solar eclipse on December 14, 2020. The December 2020 total solar eclipse has a maximum duration of 130 seconds (two minutes, 10 seconds).

There aren't many cities or large towns in the region of this solar eclipse, with Temuco (pop. 275,000), Chile, being the primary city in the path of totality. Other Chilean towns in the path of totality include Villarrica (pop. 49,000) and Pucón (pop. 22,000). These communities will likely have a surge of tourism surrounding the total solar eclipse, so it's best to plan ahead if you hope to visit at that time. Parque Nacional Villarrica, home to three volcanoes, is in the centre of the eclipse path and has camping available at Quetrupillán if you want to experience the spectacle in the thick of an evergreen araucaria forest.

December marks the start of Chile's summer, providing a great excuse to take that bucket-list trip down the coast of Patagonia in a Navimag ferry to Puerto Natales along the fjords; the ferry departs from Puerto Montt, five hours south of Temuco by bus. Beware, however, if you're trying to combine your solar eclipse experience with a trip to the Atacama Desert's famed observatories: the two regions are at opposite ends of this long, narrow country. Before flying back home from Santiago, ring in a festive New Year's Eve in Valparaíso on Chile's coast.

While totality passes over only a small section of the South American continent, the majority of residents as far north as Ecuador, Peru and Brazil will be able to see a partial eclipse. Cities including Lima, Peru; Santiago, Chile; Rio de Janeiro, Brazil; and Buenos Aires, Argentina, will all experience the partial solar eclipse. The partial eclipse will also be visible in parts of Antarctica and Southern Africa; you could even see the partial eclipse from Cape Town, South Africa, or Windhoek, Namibia. Be sure to practice safe eclipse viewing in the partial eclipse path to avoid eye damage. Pack your own eclipse glasses or make a viewer yourself.

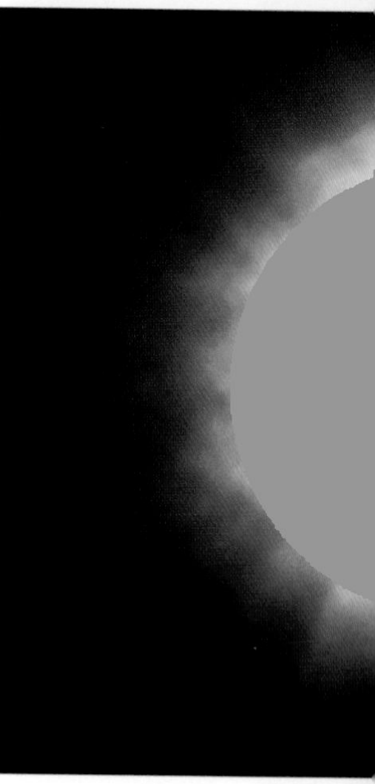

Clockwise from top: The region's distinctive araucaria tree; on the street in Barrio Italia, Santiago; a total solar eclipse with Baily's beads.

© Francisco Negroni / Alamy Stock Photo

If travelling to the region for the total solar eclipse, consider beginning your trip two weeks earlier. On November 30, 2020, a penumbral lunar eclipse will pass over the region. Unlike during a total or partial lunar eclipse, the moon won't be fully in the Earth's shadow during this eclipse; instead, a light shadow will appear over the moon. While this type of eclipse is more subtle than a blood-moon lunar eclipse, it's still a rare chance to see the moon change its appearance due to the shadow of our planet.

Important Info

Eclipse details: The eclipse maximum is at 16:13:28 UTC (Coordinated Universal Time, previously known as GMT) on December 14, 2020.

Top cities in the path of totality: Temuco, Villarrica and Pucón, all in Chile.

Villarrica volcano.

Antarctica

2021

© David Merron Photography / Getty Images

A group of Adélie penguins rest on vivid blue ice.

Most travellers visit Antarctica to see penguins and explore one of the last great frontiers of the world; for space enthusiasts, there will also be a chance to experience a total solar eclipse here on December 4, 2021. The December 2021 total solar eclipse will pass over Antarctica with a maximum duration of 231 seconds (three minutes, 51 seconds), but unfortunately it will be virtually impossible to experience this duration due to the path of totality and the few sites of human settlement in the region. Most base camps are in the Antarctic Peninsula, which is encircled but not crossed by the path of totality. The bulk of the path instead crosses close to the geographic South Pole or in the wild seas of the Southern Ocean. Forbidden fruit, indeed. If you do go, it's less of a typical vacation and more of an all-out expedition.

Given that totality is passing over one of the most isolated parts of the world, there is only one permanent settlement within the path of the eclipse: Orcadas, Antarctica (pop. 45). The eclipse will reach Orcadas Base near its end, so there will be only 58 seconds of totality here. Otherwise the show will be displayed to a large population of penguins. Since it's taking place over the icy expanse of glaciers on this most remote continent, the eclipse is sure to be an even more hushed and awe-filled experience than most – and not just because so few people will be in sight to disturb the quiet.

If you're set on seeing this remote eclipse, consider booking a cruise to the region during this time. Ship and tour operators will attempt to pass through the path of totality, if possible, as part of more traditional Antarctic expeditions. As this is the prime season for visiting the Antarctic continent, it could be a good chance to combine several once-in-a-lifetime travel experiences. Even more extreme (and pricey) is the option to take an eclipse flight out of Punta Arenas, Chile. The ultra-swell might go so far as to stay in the private Union Glacier Camp, which maintains its own airstrip. The confluence of a total eclipse in Antarctica is sure to give rise to some epic travel planning. With the land of the midnight sun experiencing the drama of maximum eclipse, interest will surely be high despite limited access.

The path of the partial eclipse is more accommodating to travellers. Major cities that will see a partial eclipse include Cape Town, South Africa; Queenstown, New Zealand; Canberra and Melbourne, Australia; and Hobart on Tasmania.

From left: The old church at the former whaling town of Grytviken in South Georgia; cruise ship, Lemaire Channel.

© BigRoloImages / Shutterstock; © evenfh / Shutterstock

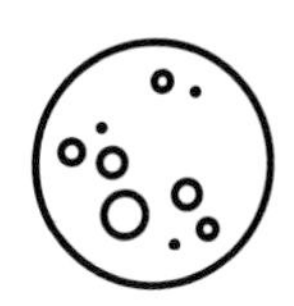

Many Antarctic cruise itineraries now include stops at the Falkland Islands or South Georgia and the South Sandwich Islands. These formerly uninhabited island groups were discovered, colonised, and settled by Europeans in the 17th and 18th centuries. Both island groups are now home to military bases that conduct research and conservation efforts, as well as the justly famous penguin colonies. It's possible to experience the southern lights during the shoulder season, but travel is limited by the pack ice.

Important Info

Eclipse details: The eclipse maximum is at 07:33:26 UTC on December 4, 2021.

Top cities in the path of totality: None. The only permanent settlement in the path of totality is Orcadas Base on the South Orkney Islands.

Oceania and Maritime Southeast Asia

2023

On the map, snaking across Southeast Asia, the moon's shadow will plunge the region into darkness ... but almost as if by design, the April 20, 2023 eclipse will miss any major cities or towns as it passes across Australia, East Timor, and Indonesia's West Papua. Because the eclipse's maximum duration is a short 76 seconds (one minute, 16 seconds), only the most dedicated eclipse chasers will make the trek to experience this total solar eclipse. Those who do may perhaps be drawn by the relative rareness of a hybrid eclipse where, depending on location, viewers will experience either a total solar eclipse or an annular eclipse. If you count yourself among those travellers up for extra adventure in order to see the eclipse, here's what you'll need to know.

The path of totality for the April 2023 total solar eclipse manages to snake from the edge of the Australian mainland across the scattered islands to its northeast, ranging from the Indian Ocean to the western Pacific Ocean. While there are certainly small communities on the island nation of East Timor and on West Papua in Indonesia that will experience this brief eclipse in its full glory, many of these will probably be unable to bear the burden of large-scale tourism; they lack the amenities many eclipse chasers might expect. West Papua, Indonesia's largest and most easternmost province, is likely to be of interest mostly to intrepid explorers and amateur ethnologists, and even

Aerial of Turquoise Bay, Exmouth, Australia.

those might tread lightly. The eclipse's path falls across a largely rural swathe of the region.

One of the largest communities in the path is Exmouth, Australia (pop. 2200); it will likely see a surge of tourism, so make your plans early. Camping out in the protected wilderness of Cape Range National Park is one good option for the outdoors-inclined, as long as online reservations are made well in advance. Even more luxe? Consider the option to charter a liveaboard boat into the path of the eclipse along Ningaloo Reef, a World Heritage Site. Australia's longest fringing coral reef is located just off the coast and is one of the world's prime spots to view whale sharks. The April hybrid eclipse falls conveniently within the best season for seeing the gentle giants in these

From left: Whale shark and hangers-on; the Milky Way above the Henbury Craters in Australia.

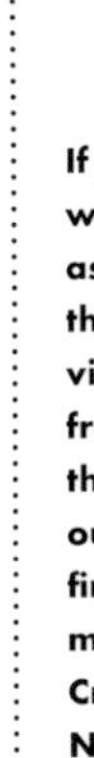

© Rich Carey / Shutterstock; © National Geographic Image Collection / Alamy Stock Photo

waters. East Timor is also ringed by coral reefs and vibrant marine life, making these eclipse destinations a good one-two for any astrotourists with a scuba certification or a love of snorkeling.

Do East Timor, West Papua and Western Australia feel too remote and inaccessible? Some major cities that will experience at least a partial solar eclipse in April 2023 include the much more central cities of Perth, Australia; Jakarta, Indonesia; Manila, Philippines; Singapore; and Ho Chi Minh City, Vietnam. In fact the whole region will experience at least a partial eclipse, including Malaysia, Taiwan, parts of southern Japan, and all of Papua New Guinea, Indonesia and Australia. Many of the larger cities in this region will likely host eclipse events and tours.

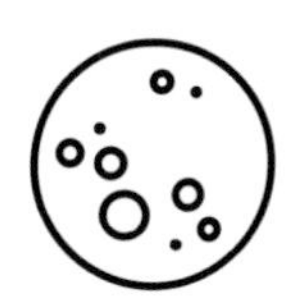

If you're eager to get well off the beaten path as part of experiencing this eclipse, consider viewing the eclipse from Exmouth and then setting off into the outback. There you'll find three fantastic meteor craters: Wolfe Creek Meteorite Crater National Park in Western Australia, and Gosses Bluff crater and Henbury Meteorites Conservation Reserve in the Northern Territory. You'll need a car to visit all three.

Important Info

Eclipse details: The eclipse maximum is at 04:16:49 UTC on April 20, 2023.

Top cities in the path of totality: None. Instead, consider planning your trip to enjoy the partial eclipse.

North America

2024

© Witold Skrypczak / Alamy Stock Photo

Mule Ears Peaks at sunset, Chihuahuan Desert in Big Bend National Park, Texas.

Almost as soon as totality ended during the August 2017 'Great American Eclipse', people began talking about when the next total solar eclipse would occur in the region. Luckily for the millions of travellers who journeyed to the path of totality (and those who regret not doing so), it won't be a long wait: the next total solar eclipse will pass across North America on April 8, 2024. This eclipse will also last much longer than in 2017, at 268 seconds (four minutes, 28 seconds). While the eclipse won't last that long across the entirety of North America, it will likely feel substantially longer than the previous one to those who experienced totality in 2017.

Passing in a northeastern direction across North America, this solar eclipse will be visible throughout the entire continent – even including parts of southeastern Alaska! As with the August 2017 eclipse, it's expected that several million travellers will make their way to the path of totality in April 2024. Because of this, make plans well ahead if you want to travel for totality; hotels and tour operators will likely be booked up months in advance, and highways and roads will be very busy. Plan for extensive travel delays if you're travelling on the days surrounding the eclipse; consider making a longer trip and visiting the city or nearby dark-sky attractions in the area.

A number of major cities are in the path of totality. These include Mazatlán and Durango, Mexico; Dallas and Austin, Texas; Indianapolis, Indiana; Cleveland, Ohio; and Montréal, Québec. Along the path many smaller towns and communities will experience totality as well.

Even some Dark Sky Parks are within the path of totality, ideal additions to your trip itinerary. These include Big Bend National Park on the Texas–Mexico border; Enchanted Rock near Austin, Texas; Geauga Observatory Park in Montville, Ohio; and Mont-Mégantic in Québec (a Dark Sky Reserve, p48). View the eclipse here and you can add on night-sky activities such as stargazing.

Outside the band of totality, major cities will experience a partial solar eclipse, including Mexico City, Toronto, New York City, Chicago, Los Angeles and Washington, DC. Parts of the South Pacific, Central America, and even continental Europe will experience a partial solar eclipse too, making this likely to be among the most widely viewed eclipses ever!

From left: Mazatlán beach in Mexico; Dallas skyline at sunrise.

© photomatz / Shutterstock; © wsfurlan / Getty Images

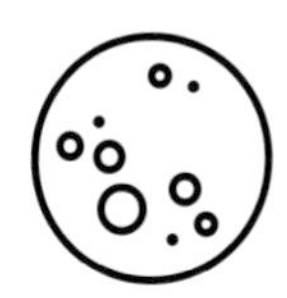

Indianapolis, Indiana, lies directly in the path of totality for the 2024 solar eclipse. Known as the Crossroads of America, this city will draw eclipse chasers from all around, and it's ready for the crowds. Home to the Indianapolis 500 (aka the Indy 500), the Indianapolis Motor Speedway plans to host up to 350,000 people for 'The Greatest Spectacle in Eclipse Watching'.

Important Info

Eclipse details: The eclipse maximum will occur at 18:17:16 UTC on April 8, 2024.

Top cities in the path of totality: Maximum duration will occur in the town of Nazas, Mexico (pop. 3600). Major cities like Austin and Dallas in Texas are also great choices due to substantial tourism infrastructure to support the number of eclipse chasers.

Europe

2026

Europeans may wonder when their turn to experience a total solar eclipse is coming, after all the excitement about eclipses in North America. August 2026 will be their chance – though travellers from around the world will likely flock to Europe for this eclipse too, thanks to the path it cuts across some of Europe's most popular destinations. On August 12, 2026 as the sun goes down, the shadow of the moon will pass across the Earth, with a maximum duration of 138 seconds (two minutes, 18 seconds). While most major cities in the path of totality won't experience an eclipse this long, you can plan ahead to experience as much as possible of this major event, or opt to watch the partial eclipse from another city such as London, Paris or Rome.

While Greenland isn't as easily reached as other destinations where you can see the 2026 solar eclipse, it's the most otherworldly – and a great destination for adventurous astrotourists who also want to try other Arctic activities after the roughly two minutes of totality that East Greenland will experience for this eclipse. Plan ahead to book a tour or cruise of the fjords and glaciers of East Greenland, and try your hand at dog sledding, snowshoeing or skiing. By mid-August when this eclipse occurs, you can also spot the northern lights once the sun goes down.

Iceland is perhaps the best and most accessible destination for experiencing the 2026 solar eclipse in totality: daily flights from North America and Europe make it easy to reach Reykjavík, which is the largest city near the point of greatest duration in the eclipse. From Reykjavík or in nearby Thingvellir National Park, you can experience a brief 63 seconds of totality before the two-hour-long partial eclipse continues its second half. The sun will be sinking towards the horizon throughout the eclipse, so wherever you choose, be sure it has a good view of the western horizon. As with viewing the eclipse in Greenland, you could potentially tack on an early-season aurora-viewing trip too.

On the European continent, the path of totality cuts a wide swathe across the Iberian Peninsula, and much of Spain will get to enjoy the

Clockwise from top: The northern lights over Oqaatsut village in West Greenland; dogs pulling a sled through snow in East Greenland; a sunset over Strokkur geyser in Iceland.

© Stewart Smith / Alamy Stock Photo ; © Juanma Aparicio / Alamy Stock Photo

full experience; elsewhere in Europe and parts of North Africa, viewers will experience a partial eclipse lasting roughly between 45 minutes and two hours as the sun sets. To experience totality in Spain, travel to cities like Bilbao on the northern coast, Valladolid in central Spain, or Valencia on the shores of the Mediterranean. In most cities totality will occur with the sun mere degrees above the horizon; you will definitely need a clear view to the northwest in order to experience the total eclipse. Totality is also quite brief, ranging between 36 seconds in Bilbao to 89 seconds in Valladolid. Madrid and Barcelona are just outside the path of totality, but you can base yourself in either city and travel on the day of the eclipse itself if you get an early start.

From left: View of Reykjavík in Iceland; the Guggenheim Museum, Bilbao.

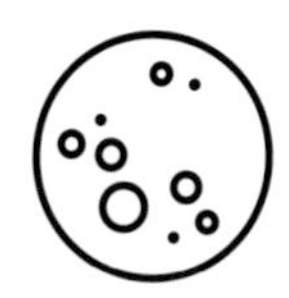

No matter which European city you choose to base yourself in for the 2026 eclipse, there are likely more astrotourism experiences nearby. In Greenland it may be possible to see the aurora on the nights surrounding the eclipse; while in Iceland, take a trip to Jökulsárlón and keep your eyes peeled for the northern lights too; and in Spain, plan a side trip to Albanyá.

Important Info

Eclipse details: The eclipse maximum will occur at 17:46:01 UTC on August 12, 2026.

Top cities in the path of totality: Maximum duration will occur over the North Atlantic Ocean east of Reykjavík, Iceland. In Spain, Bilbao, Valencia and Valladolid are all in the path of totality. Ibiza in Spain will also experience a brief 74 seconds, making it an option for nightlife-focused travellers.

North Africa

2027

Imagine the sun disappearing from the sky over the Nile or between the European and North African cities that line the Strait of Gibraltar. On August 2, 2027, a total solar eclipse will plunge North Africa and small areas of Europe and the Middle East into darkness. The moon's shadow will pass along the northern coast of the African continent, traverse the Red Sea and part of the Arabian Peninsula, and even cross the Horn of Africa. This eclipse will last an astonishing 383 seconds (six minutes, 23 seconds), making it the second-longest total solar eclipse in the 21st century.

Travelling to see the 2027 solar eclipse will take advance planning, though the North African region continues to develop infrastructure for tourists. In particular, consider countries that have strong tourist economies in the years leading up to this eclipse. Southern Spain (including the cities of Málaga and Cádiz) will be a good destination; so will cities such as Tangier in Morocco and communities along the Nile in Egypt. Egypt's Asyut will be the town closest to the point of maximum duration, experiencing over six minutes of totality and nearly three hours of the partial eclipse. Luxor of pyramid fame will experience totality as well, while the Nile port city Aswan will be just outside the totality zone.

Parts of northern Algeria and Tunisia will also experience totality, as well as northeast Libya. Local and global tour operators will likely have a wide range of offerings that will allow you to experience desert activities in the northern Sahara along with viewing the solar eclipse. Saudi Arabia may be a good destination for some travellers, as the city of Jeddah falls directly in the path of totality; the Yemeni capital city of Sana'a will experience totality too. Keep an eye on political tensions and upheaval in the region as you set up your eclipse trip. Lastly, the northeast tip of Somalia, including the city of Bosaso, will also experience totality. These offbeat destinations may call out to adventure-seeking astrotourists.

Major cities in the region, including Rabat, Morocco; Algiers, Algeria; Tunis, Tunisia; and Cairo, Egypt, are all close to the path of

Málaga in Andalusia will be in the eclipse path.

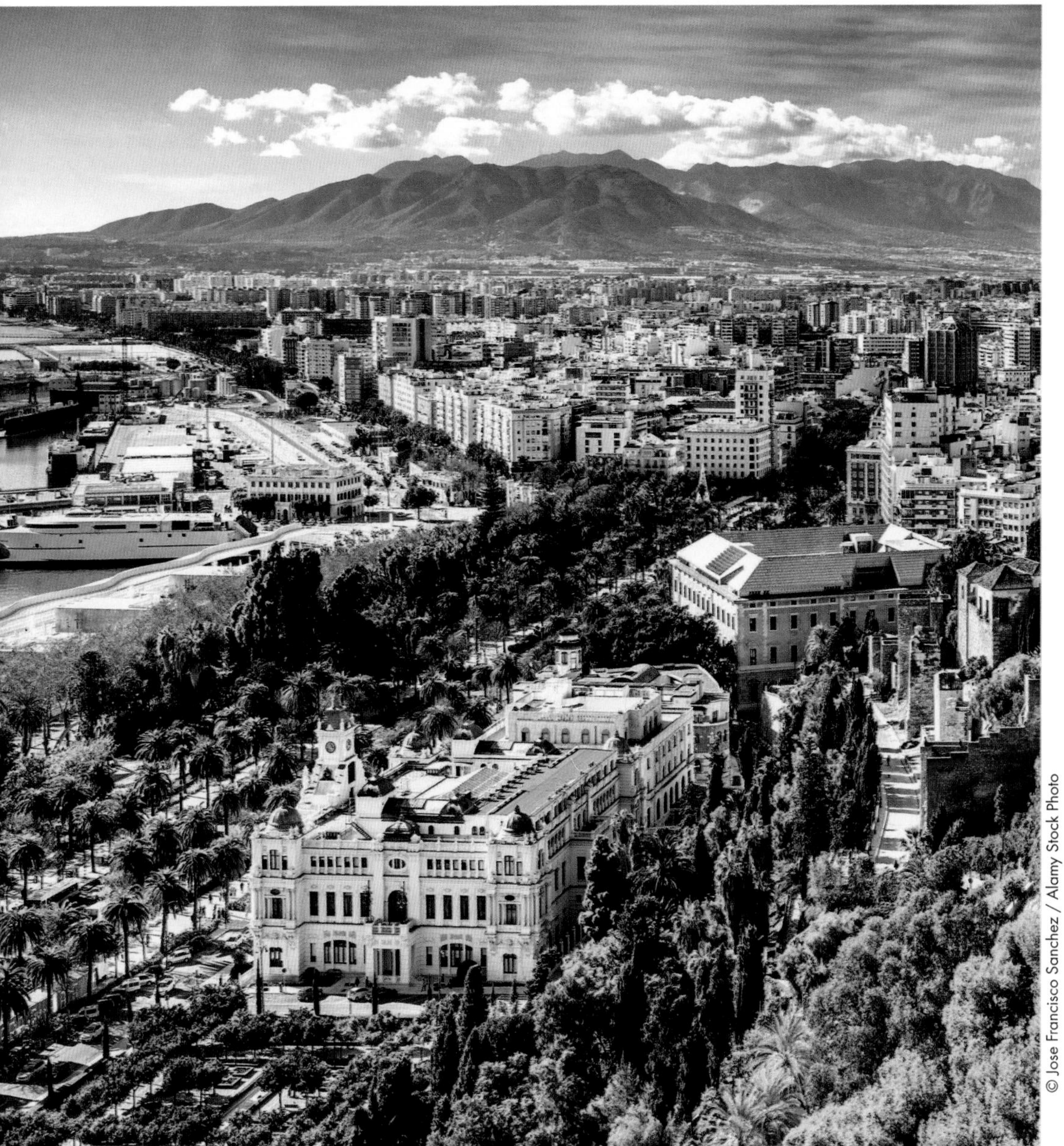

© Jose Francisco Sanchez / Alamy Stock Photo

totality, but not quite within it. If you base yourself in one of these cities, plan to travel into the path of totality for an overnight or multi-day trip. Tour operators will likely offer single- and multi-day tours to take advantage of the tourism boom the eclipse will undoubtedly inspire.

© Evenfh / Shutterstock

If you're going to visit Morocco for the solar eclipse, consider adding on a few nights near Merzouga and Erg Chebbi (p72). This is one of the many great spots for stargazing in eastern Morocco, and a number of tour providers offer guided observing tours and desert stays where you can enjoy the starry skies all night long.

Important Info

Eclipse details: The eclipse maximum will occur at 10:06:36 UTC on August 2, 2027.

Top cities in the path of totality: Málaga and Cádiz, Spain; Tangier, Morocco; Luxor, Egypt; Jeddah and Mecca, Saudi Arabia; and Sana'a, Yemen, are all good options for travellers making the effort to see this historically long eclipse, political conditions permitting.

Dunes of Morocco's Erg Chebbi Desert.

Australia and New Zealand

2028

While only the tiniest sliver of Australia will experience the total solar eclipse in 2023, a much larger portion of the country, including Sydney, will be in the path of totality on July 22, 2028. This total eclipse will last a maximum of 310 seconds (five minutes, 10 seconds), with totality passing through the Australian Northern Territory and the states of Western Australia, Queensland, South Australia and New South Wales. Totality will also cross New Zealand's South Island.

© dave stamboulis / Alamy Stock Photo; © Taras Vyshnya / Shutterstock

From left: Tunnel Beach, New Zealand; Sydney CBD and harbour.

For the most part, the 2028 total solar eclipse passes across the Australian outback, a relatively remote and undeveloped part of the country. Adventurous travellers may want to pack up a van to head out on their own, or book with a tour provider heading into the region to experience the eclipse from that part of the country. If you're chasing the eclipse for the longest possible duration of totality, head for the northern part of Western Australia, where you can experience the full 310 seconds near Mitchell River National Park.

For city-loving travellers, there's only one place to be: Sydney, where you can experience three minutes, 58 seconds of totality as part of a 2½-hour eclipse. The capital city of New South Wales and the surrounding towns along the shores of the Tasman Sea will undoubtedly host eclipse events and tours to the region. As this eclipse occurs in Australia's winter, there's a roughly 50% chance of cloud cover. If the skies are clear, look for the eclipse in the northern sky as the sun moves towards the horizon; you'll need a view of the northwest horizon for the best chances of seeing it. Consider tacking on a trip to Tasmania (p203) to see the southern lights as part of your trip down under.

Unfortunately, Uluru (Ayers Rock) will not experience totality during the 2028 eclipse. Instead, you can experience an 80% partial solar eclipse at this famous Australian site.

In New Zealand, totality will cut across the South Island. Whether you're viewing from the fjord lands on the western coast or the city of Dunedin in the east, totality in New Zealand will last roughly two minutes, 55 seconds as part of the over-two-hour partial eclipse. You can also extend your trip to New Zealand to chase the southern lights, primarily visible from the southernmost parts of the South Island. Aoraki Mackenzie Dark Sky Reserve (p78) is another worthy site.

From left: Kalbarri National Park coast; Grand High Tops in Warrumbungle.

© David Noton Photography / Alamy Stock Photo; © Ingo Oeland / Alamy Stock Photo

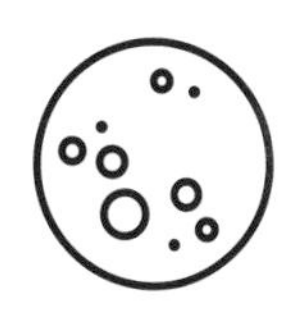

Warrumbungle National Park (p40) is one of the best places in Australia to enjoy dark skies; it was the first designated Dark Sky Park in the country. Warrumbungle is also a great place to watch the solar eclipse in 2028, as it is in the path of totality. Make a full space-themed trip by spending time at Siding Spring Observatory during your visit.

Important Info

Eclipse details: The eclipse maximum will occur at 02:55:22 UTC on July 22, 2028.

Major cities in the path of totality: Sydney, Australia and Dunedin, New Zealand are the two biggest cities along the path of totality. Consider smaller towns in the surrounding area if you want to avoid the biggest crowds.

Launches

Rocketry dates back to the 10th-century Song dynasty in China. Originally used to propel arrows, rockets have, for most of history, been weapons of war. Only in the 20th century have rockets been used to advance different endeavours of humanity: technological innovation, scientific research and even space tourism. Beginning in 1957 with the launch of Sputnik, humankind has put objects into space at an astonishing rate. Now almost 2000 satellites orbit the Earth, dozens of space probes are exploring the far reaches of our solar system – and all of them were put in space with a rocket.

To watch a rocket rise from a launch pad into the sky, defying that one constant of physics, gravity, is to witness an amazing feat of human innovation. It's an unforgettable experience for any who have seen it. The loud roar of the boosters reverberates deep in your chest, and from miles away you can feel the heat as the rocket takes flight. Even though you're viewing from a distance, the thrill makes launch viewing one of the most adrenaline-producing things to observe.

While there have been nearly 40,000 rocket launches in the first six decades of space flight, most people have never experienced one. This is in part because the location of each launch site is typically remote, away from human settlement. Launching to space is a high-risk business, and for decades the process included dropping massive portions of the rocket and boosters back to Earth, usually landing in the ocean. To minimise the impact of launches, facilities were built away from cities and near the coast. Over time, cities and towns have developed closer to spaceports, and it has become easier to travel to these locations.

If you'd like to see a rocket launch for yourself, flexibility is the name of the game. While rocket launches are spectacular to view, they are missions, not tourist events. This means that their schedules are variable; launches are often delayed and rescheduled at the last minute because of technical problems or weather. Dates are seldom set more than a few weeks in advance (unless it's a really big launch), which means that making plans to see them can be difficult. Spaceflight Now has an excellent searchable list of upcoming launches from around the world, with a brief description of each. If you have a choice between a crewed and uncrewed launch and scheduling is a concern, always opt in favour of a launch without astronauts. There are fewer factors to consider in terms of launch conditions, and that makes it more likely that the rocket will launch on time (but again, it's good to prepare for delays regardless of the type of launch). Additionally, avoid demo launches. While a new type of rocket may take off on its first scheduled date,the odds are higher that things will go wrong during a test launch.

A limited number of countries have active launch facilities, and even fewer have sites open to the public and foreign travellers. If you're willing to travel and able to have a flexible itinerary to accommodate launch delays, it's possible to watch a launch from the Florida coast, near the Amazon rainforests of French Guiana, a remote peninsula in New Zealand and even the bustling streets of a city in southern India. Launches continue to increase in frequency around the world, making it possible to experience this unforgettable feat of human engineering for yourself. Yes, planning for a launch can be tricky, but once you are there and the energy begins to build as the clock counts down, you'll know that it was worth it. Along the way, each destination offers its own range of other attractions, whether space-related or not.

China

CNSA

In 2018 China launched as many government-funded rockets as the entire US aerospace industry combined, proving that it is making quick strides to become dominant in the industry – and in space. China was the third country to send a crewed vessel into space, including a space station of its own.

As a tourist destination China already draws interest for its history, culture and food, and astrotourists will see more opportunities to experience rocket launches and other space-related events in the coming decades. The four major launch centres in China vary in their frequency of launches and public accessibility (especially for foreigners), but it's certainly possible to witness a launch if you're already travelling in a region of China when one is happening.

Jiuquan Satellite Launch Center, also known as Dongfeng Aerospace City, is in the Gobi Desert of northern China in the region known as Inner

A camel train in the dunes of the Gobi Desert.

Mongolia. Established in 1958, Jiuquan is the oldest and longest-operational launch facility in China, and was the primary site used for Chinese human space-flight launches, including to the now de-orbited Tiangong-1 Chinese space station. Home to nearly 1 million people, the nearby town of Jiuquan is a base for those who want to explore dynastic Chinese history. Several notable sites including the Weijin Tomb murals from the Jin Dynasty, the ancient Mogao Caves and the Wine Spring of the Western Han Dynasty are in the area. Combine a visit to these sites with an attempt to watch a launch to better appreciate the past, present and future of life in China.

Taiyuan Satellite Launch Center is located west of Beijing in Shanxi province. This facility opened in 1968 and has been active since, launching scientific satellites, imaging and weather satellites. It's possible to stop in the area of Taiyuan, which is a 6½-hour drive from Beijing, while visiting other cities in this part of North China, including Shijiazhuang with its temples, monasteries and mountain hot springs. You can also take the train from Beijing to Shijiazhuang and drive to Taiyuan if a launch is expected. While in Beijing, check out the China Academy of Launch Vehicle Technology (CALT) museum along with visiting the Great Wall and Forbidden City.

Xichang Satellite Launch Center is the southernmost active Chinese launch facility, located south of Chengdu in Sichuan province. Since it opened in 1984, the facility has been used to launch geostationary navigational and communications satellites, as well as other surveillance and Earth-imaging satellites. In 2007 it was proposed that Xichang Satellite Launch Center would open for tourists to visit during launches, but the facility does not allow foreigners and public launch-viewing opportunities are limited.

Perhaps the most public-facing of the Chinese launch facilities, Wenchang Spacecraft Launch Site is on the island of Hainan in southern China, otherwise known for its enticing tropical beaches. It is the newest launch facility, opened in 2014 to take advantage of better geography for increasing launch payloads; since then, there have been a handful of launches from Wenchang to test manned space-flight technology for China's next generation of rockets and capsules. Wencheng is a 60- to 75-minute flight plus an hour-long drive from Hong Kong, Nanning, and Guangzhou, making it one of the easiest launch facilities to visit in China. There has already been a surge in tourism to watch launches from Wenchang on Hainan's beaches. If you're planning to visit China and want to see a launch, you can travel independently to Wenchang to experience one; a 'tourism town' is planned to encourage visitors.

What's in a name? While Americans are called 'astronauts', Russians refer to these space explorers as 'cosmonauts', and the French call them 'spationauts', Chinese space travellers are called taikonauts. This term was developed in the late 1990s – combining the Mandarin characters for space, 太空, with the suffix 'naut' – by a Malaysian news reporter!

Important Info

Launch info: China is tight-lipped about specific launch windows and payloads, often launching with no international announcement and little domestic notice. Some websites do list launch information about upcoming Chinese launches, but it's best to assume these are general guides. Plan to be in the area for up to a week if you want to see a launch – some will launch 'early', and others will launch 'late'.

French Guiana

ESA

Located on the northern coast of South America, French Guiana is at first glance a surprising place for the European Space Agency (ESA), one of the world's largest space agencies, to choose for a launch location. Until, that is, you consider that French Guiana is ideally placed within a few degrees of latitude from the equator, allowing launches of larger payloads with less fuel, and that as its name suggests, it is a department of France, which is home to ESA's headquarters.

The remote location and mostly undeveloped rainforest and coastline of French Guiana are ideal for launching with minimal risk to human settlements, making this destination perfect for adventurous astrotourists willing to go way off the beaten path to explore one of the least-visited places in the world. With its unique mix of Caribbean, American and European culture, French Guiana makes an ESA rocket launch a truly multicultural affair.

ESA was established later than many space agencies, in the mid-1970s. Supported by the EU as well as individual member countries, the agency oversees all European efforts to launch satellite payloads as well as astronaut training for European astronauts who work aboard the International Space Station. Through the late 2010s, European astronauts have historically launched with American and Russian ISS crewmates from Baikonur Cosmodrome in Kazakhstan rather than from French Guiana, which specialises in satellite payload launches.

The Guiana Space Centre, or Centre Spatial Guyanais (CSG), is located outside the town of Kourou, an hour's drive north along the coast from Cayenne, the capital of French Guiana. Home to 25,000 people and the CSG Space Museum, Kourou is the best place to base yourself if visiting French Guiana for a launch. Here visitors can learn more about the history of ESA and launches from the site. On launch days there are a variety of public viewing locations, but visitors require written permission to be allowed on the CSG grounds. This can be obtained via email to the

Launchers at Guiana Space Centre in Kourou, French Guiana.

© HomoCosmicos / Alamy Stock Photo

launch company in advance.

While the CSG is not as active as other launch facilities around the world, the ESA is reliable in its launches and publicises them in advance. Since most travellers might not find themselves in French Guiana by chance, plan a flexible itinerary to allow for launch delays if you're determined to see an ESA launch; luckily, the location near the coast is an easy place to while away the time. Visitors can take a tour of the local savannah to learn about the flora and fauna that cohabitate with rockets and astrophysicists.

Joe Pappalardo has visited most of the spaceports around the world, as documented in his book *Spaceport Earth*. **While French Guiana may not be as established for hosting tourists as other launch countries, it's still an exciting trip. 'Bring binoculars even to a night launch', he advises. 'The sky is so dark you can really see the stages separate at high altitudes, and the falling boosters seem to flicker as the red-hot engine tip goes end over end'.**

Important Info

Launch info: Arianespace, the primary launch contractor operating at CSG, publicises launch information about non-confidential payloads on their website in advance. Obtain written permission in advance to view from the spaceport.

Tour operators: The National Centre for Space Studies (CNES) has guided tours most days at CSG.

India

ISRO

© Jayakumar / Shutterstock

The Hindu Kapaleeshwarar Temple in Chennai.

Most travellers visit India to experience its culture, try local cuisine and explore historic sites; astrotourists can also add the option of seeing a rocket launch to the itinerary. While India's launch schedule requires flexibility, sometimes on the order of weeks or months of delays, with patience it's possible to see one of the Indian Space Research Organisation (ISRO) launches off the country's eastern coast, north of Chennai.

The ISRO was established under a different name in 1962 and renamed in 1969. The Indian response to the global space race, ISRO has developed technology comparable to that of more established countries and space organisations like NASA in the US, Roscosmos in Russia and ESA in Europe. Today ISRO primarily facilitates Indian satellite launches and the development of technology to support those payloads.

Satish Dhawan Space Centre, in the state of Andhra Pradesh, is the primary ISRO launch facility and the only one from which launches have occurred in the past several years. Foreign visitors are not permitted to tour the facility even on non-launch days, but the public may be able to view launches from nearby Pulicat Lake (which turns into a salt lakebed during the dry season) and the surrounding shore, depending on the season and time of launch. The lake is also a bird sanctuary.

India has long wanted to join the ranks of countries putting humans into space. In 2018 Indian Prime Minister Narendra Modi declared that the Indian Human Spaceflight Programme (IHSP) will aim to send its first crewed mission into space by 2022.

Important Info

Launch info: The ISRO does publish information on launch windows and non-classified payloads, but these dates are often pushed out due to technical and weather delays. Plan ahead and give yourself extra time in the area if you have your heart set on seeing a launch.

Tour operators: There are no known tour operators offering guided tours to see ISRO launches as yet.

Japan

JAXA

Japan has proven itself a leading global player in the aerospace industry, and in the 21st century is advancing its own launch capabilities and innovating in other areas of space technology. Relatively small and blessed with excellent transit links compared to other countries with launch capability, Japan is extremely welcoming to tourists, including foreigners, who visit launch facilities on non-launch days. If you're interested in combining a trip to futuristic Tokyo with an experience of Japan's rocket launches, some advance planning and a bit of off-the-beaten-path travel can make that possible.

Both of Japan's launch facilities are in the far southern part of the country: Uchinoura Space Center is located in the Kagoshima prefecture on the island of Kyushu, and Tanegashima Space Center is located even further south on the island of Tanegashima. Both are situated to take advantage of the southernmost geographical positions from which Japan can launch, using the Earth's rotation to increase fuel efficiency and payload sizes.

If you plan to visit Fukuoka or Nagasaki on Kyushu (Nagasaki has an excellent atomic bomb memorial), it's easy to make a loop of the island and visit Uchinoura Space Center too. More public-facing and easily accessible than Tanegashima, Uchinoura is open to the public on non-launch days for tours of the facilities; it has exhibits curated by the Japanese Aerospace Exploration Agency (JAXA). On launch days it's even possible to view the action from an observation deck.

To visit Tanegashima Space Center, you can catch a ferry from Kagoshima to the island of Tanegashima and drive to the space centre from the ferry terminal. Tanegashima is home to a space museum, including a full-size replica of the Japanese module of the International Space Station, and is open for tours daily. While you can't watch launches from Tanegashima itself, they can be viewed from the surrounding area, including designated viewpoints like Rocket Hill Observatory, Takesaki Observation Stand and Kamori-no-mine.

Clockwise from top: Tsukuba Space Center; the futuristic Shinkansen train; launchers at Tanegashima Space Center.

There are Japanese space centres and facilities across the country, and many are open to the public. If you're visiting Tokyo, add Tsukuba Space Center (TKSC) to your itinerary; this facility is the Japanese equivalent of NASA's Johnson Space Center. And don't forget to look into the logistics of visiting Japan's first certified Dark Sky Park, Iriomote-Ishigaki National Park in Okinawa.

Important Info

Launch info: JAXA is good about publicising its launch windows, but the dates are often pushed back or scrubbed. If you plan to travel to Japan to see a launch, be flexible in your timing and add on other activities in the area.

Tour operators: You can visit Uchinoura and Tanegashima independently; no known tour operators go to either facility at present.

Kazakhstan

ROSCOSMOS

© NASA Photo / Alamy Stock Photo

A rocket launches from Baikonur Cosmodrome.

If you want to see a launch by the Russian space agency Roscosmos, then your best (though seemingly counterintuitive) bet is to visit Baikonur Cosmodrome in neighbouring Kazakhstan. This Central Asian country is the ninth largest in the world, and the largest landlocked country.

Originally established when Kazakhstan was part of the Soviet Union, Baikonur Cosmodrome is the oldest and largest operational launch facility in the world. Baikonur is where Sputnik was launched, and where Yuri Gagarin boarded Vostok 1 to make his historic launch to be the first man in space. Through the early 21st century, it was used as the primary Soviet/ Russian launch facility for human space flight, including crew launches to the International Space Station.

While it's possible to see a launch in the area, there are no civilian airports near Baikonur Cosmodrome, so you'll need to travel overland to visit this site on the Kazakh Steppe in a remote area of western Kazakhstan. Thus, the best way to see a launch here is by booking a prearranged tour that coincides with a launch. Several tour operators offer private tours that use chartered jets to zip tourists from Moscow to the town of Baikonur; the tours show off the famous sites of the cosmodrome as part of a multi-day itinerary. It's a unique opportunity to visit the cradle of space flight and feel the rumble of the engines pounding your chest with sound waves.

Val Chebakin is a programme manager at Space Adventures, a global space tourism company with tours to Baikonur Cosmodrome. 'The most impactful thing about visiting Baikonur is the launch itself, the sensory experience for those witnessing the launch', he says. Chebakin advises against trying to photograph a launch here: 'If you concentrate on taking pictures, you will end up with disappointing results and you would have missed the true spectacle. Just enjoy the view.'

Important Info

Launch info: ISS-related and non-classified payload launches are typically publicised well in advance; launches from Baikonur are some of the most reliable in terms of launching within the first available launch window.

Tour operators: Try Space Adventure, Russia Flight Adventures and Star City Tours.

New Zealand

ROCKET LAB

New Zealand is a dream destination for many reasons, not least that it's home to stunning dark skies in places like the Aoraki Mackenzie Dark Sky Reserve and Aotea (Great Barrier Island) Dark Sky Sanctuary (p80). It even offers the opportunity to see the southern lights during dark winter months. These experiences, plus New Zealand's renowned hospitality, culture and food, make it a top destination by day and by night.

In the past few years New Zealand has also attracted its first space-flight venture, and it's now possible to see a rocket launch from the Mahia Peninsula. In 2017 US-based company Rocket Lab established its private launch facility, Launch Complex 1, on the southern tip of the Mahia Peninsula of New Zealand's North Island.

Though it's a long, 7- to 8-hour drive from either Auckland or Wellington, the Mahia Peninsula can be an easy extension of a road trip to the region's famous Hawke's Bay wineries.

Before you book a trip, consider that Rocket Lab completed only two successful test launches in its first year, both with significant delays from the initial launch window. Additionally, the Mahia Peninsula launch facility is private and not open for public tours. While it's likely that launches will increase in both frequency and reliability – Launch Complex 1 is certified to launch up to 120 rockets per year – your best bet is to be flexible in your itinerary and open to the possibility that you won't see a launch during your trip to New Zealand for at least the next few years.

The best time for astrotourism in New Zealand is in the winter months, when dark skies guarantee better stargazing and chances of seeing the aurora. Head to Stewart Island; the shimmering auroras here are said to have given the island its Māori name, Rakiura, or 'glowing skies'.

Important Info

Launch info: Rocket Lab does announce intended launch windows online, but as noted, these have historically been unreliable as part of its testing process. No other operators launch from New Zealand at present.

Tour operators: There are currently no known tour operators for visiting Launch Complex 1 or viewing rocket launches in New Zealand.

Clockwise from top left: Taylor's Bay, Mahia Peninsula; a Hawke's Bay vineyard; Rocket Lab prepares a launch; the site and facilities.

This page: David Wall / Alamy Stock Photo; Opposite: © S Curtis / Shutterstock; Courtesy of Rocket Lab

Russia

ROSCOSMOS

At one time the pioneer in space flight, and eventually a collaborator in establishing the International Space Station (ISS), Russia continues to be one of the top countries in the world for rocket launches. Indeed, Russia has played a compelling central role in the development of human space flight. Established in the 1920s, the Soviet space programme, now called Roscosmos, was a pioneer in the development of intercontinental ballistic missiles (ICBMs), and Russia won the race to space in many categories, including first satellite (Sputnik, October 1957), first human in space (Yuri Gagarin, April 1961), first woman in space (Valentina Tereshkova, June 1963) and first spacewalk (Alexey Leonov, March 1965). Russia was also responsible for the first permanently manned space station, Mir, which was operational from 1986 to 2001 and served as a precursor to the present-day ISS. Without Russia's pushing forward the cause of human

© Georgy Golovin / Alamy Stock Photo; © Konstantin Shaklein / Alamy Stock Photo

From left: A Soyuz rocket taking off; Moscow's grandiose monument to Yuri Gagarin, the first cosmonaut.

space flight, it's likely our efforts in the 21st century would be nowhere as advanced as they have become.

Despite that fact, it's not as easy to reach launch facilities in Russia as it is to visit those in other countries. The country's geographical position has necessitated that cosmodromes – the term for a Russian spaceport – are in rural areas (and in neighbouring Kazakhstan) not easily visited without advance planning. However, if you're willing to tackle the logistics, it's very possible to see one of Russia's Soyuz rockets take flight.

Two primary cosmodromes in Russia launch on a regular basis; other launch facilities are used less frequently or are much more difficult to reach. Plesetsk Cosmodrome, near the White Sea in northern Russia, is the easier to reach, but less frequently used, of the two main cosmodromes. The nearest town is Mirny, not to be confused with a town of the same name in eastern Siberia, and it requires authorisation if you plan to stay in town overnight. Mirny is a 13-hour drive from St Petersburg or a 15-hour drive from Moscow; while there is an airport near Mirny, it is used exclusively as a logistical base for Plesetsk Cosmodrome and is not open to civilian flights. Plesetsk Cosmodrome is primarily used for ICBM and satellite launches.

Vostochny Cosmodrome is in Amur Oblast in the Russian Far East. This new cosmodrome is meant to reduce Roscosmos' dependence on the Baikonur Cosmodrome in Kazakhstan. Few launches have yet taken place at Vostochny, and there are no domestic or international flights to the region, leaving visitors to arrive by car or train. Officials have said that Vostochny will eventually open for space tourism by partnering with tour operators, allowing travellers to visit this extremely remote launch location, but for now access is difficult. Visitors may want to stick with famed Cosmonauts Alley in Moscow.

If you're planning to visit Russia and are keen on space, be sure to add Star City to your list. This facility on the outskirts of Moscow is where cosmonauts have trained since the beginning of human space flight, and you can tour the facilities and learn more about the Russian space programme here. Cosmonauts Alley and the Museum of Cosmonautics are must-see attractions in Moscow, highlighting notable cosmonauts from human space flight history.

Important Info

Launch info: Non-classified payload launches are typically publicised well in advance but are relatively infrequent from both Plesetsk and Vostochny.

Tour operators: No international tour operators offer tours to Plesetsk or Vostochny cosmodromes.

USA

NASA AND PRIVATE INDUSTRY

From left: Space shuttle *Atlantis* before its final mission in 2011; sabal palms in Kissimmee Prairie Preserve State Park, FL.

The space race of the mid-20th century established the US as one of the top countries in the world for aerospace technology. While public interest has at times waned, there's still a ton of activity throughout the country, and the number of rocket launches each year continues to grow as private companies enter the market.

To experience a rocket launch, you'll need to plan ahead but also be flexible since launch conditions change frequently. Throughout the US there are opportunities to see a launch, especially on the coasts, and new spaceports keep popping up across the rest of the country.

Florida's Space Coast

Long nicknamed America's 'Space Coast', the Atlantic coast of Florida has taken up the moniker with enthusiasm. Florida's coastline was identified as an ideal launch location in the earliest days of NASA, and two major facilities were built to help launch a man into orbit and

© Ben Cooper / Getty Images; © Don Johnston_SU / Alamy Stock Photo

then to the moon.

The more familiar – and more visitor friendly – facility is Kennedy Space Center, named for the 35th US president, John F Kennedy, who enthusiastically supported US efforts to explore space. The centre is home to three launch complexes, several notable NASA facilities including the Vehicle Assembly Building (VAB), and the Kennedy Space Center Visitor Center.

Watching a launch at Kennedy is relatively easy, and launch viewing can be done at several locations open to visitors throughout the facility. These include the LC-39 Observation Gantry, with its iconic views of the countdown clock, and the NASA Causeway where visitors can park and watch the launch from a safe distance. On non-launch days, rocket tourists can explore the visitor complex, wander through a rocket garden, see the space shuttle ***Atlantis*** and occasionally meet with astronauts ready to answer questions and pose for pictures.

Next door, Cape Canaveral Air Force Station is the more insular and industrial launch facility, but its proximity to Kennedy makes it possible to watch launches at Canaveral almost as easily. There are 37 launch complexes on the Cape Canaveral property, though since it's a military base it's more difficult for visitors to get an intimate look at the work NASA and private companies are doing there.

For those who travel to see a launch (and allow time for weather delays), a Space Coast trip is unforgettable.

Use Orlando, Florida, as a base for your travels to Kennedy and Canaveral. Add a dark-sky destination to your trip and make the two-hour drive south from Orlando to Kissimmee Prairie Preserve State Park. This area is designated as a Dark Sky Park and offers great stargazing opportunities.

Important Info

Launch info: Launches are public, but there are some factors to consider when choosing your free viewing location. There are multiple launch pads dotting the Space Coast, so the optimal spots are never the same and depend on where the launch pad is located. Playalinda Beach and Space View Park in Titusville and Cocoa Beach Pier are good options. There's often a lot of traffic on launch day and parking can be difficult; get to your viewing area early.

A United Launch Alliance Delta II rocket from Space Launch Complex 2, Vandenberg Air Force Base.

Central California and Southwestern US

On the West Coast of the US, it's increasingly common to see a rocket launch or spaceplane take flight. The primary launch facility is Vandenberg Air Force Base on California's Central Coast, located outside the town of Lompoc. It was primarily used for military launches until recently, but the growth of commercial companies such as SpaceX and NASA contractors like United Launch Alliance (ULA) has consistently increased the number of launches over the Pacific Ocean, as Cape Canaveral reaches capacity. As public interest in the launches has grown, the community of Lompoc has seen a rise in tourism to match; it's now common for hotels in the area to sell out on a launch weekend. Several websites run rocket launch calendars that help prospective viewers keep track of upcoming launches and plan to try to witness one. Companies like Spaceflight Now (https://spaceflightnow.com/launch-schedule) and Spaceflight Insider (www.spaceflightinsider.com/launch-schedule), and even the local tourism board (https://explorelompoc.com/events), have information on Vandenberg launches.

Lompoc and nearby Santa Barbara are situated along the Pacific Coast Highway, so you can easily reach them by driving along winding coastal California Hwy 1. Ambitious astrotourists can stop for a few days along this stretch to try to view a rocket launch before continuing one of the world's most famous road trips.

Further inland, Mojave Air & Space Port is the oldest spaceport to be authorised for spaceplanes to make a horizontal takeoff (like an airplane, rather than vertically like a rocket) as they leave Earth. Virgin Galactic and the gigantic Stratolaunch spaceplane are among those that have had successful test launches from this facility in the heart of the Mojave Desert. While the Mojave port isn't open for public tours, the facility does have monthly Plane Crazy Saturdays and an on-site restaurant.

Meanwhile, as the private space industry tries to find its wings, New Mexico's Spaceport America, near Las Cruces, exists as a tale of the deferrals often associated with these ventures. Launches have yet to take place here, but the spaceport functions as a tourist attraction nevertheless. Pair it with Roswell or Cosmic Campground, both also in New Mexico, to get a sense of the evolution of the stories told by space-age America.

Wallops Island, Virginia

Often overlooked for more bombastic launch locations in California and Florida, Wallops Island is another option for viewing launches in the eastern US. It's one of Virginia's barrier islands along the Atlantic coast, so when you travel to watch a launch at NASA's Wallops Flight Facility, you can also explore nearby Chincoteague Island and Assateague Island National Seashore to catch sight of wild ponies. Since Wallops Flight Facility was established in 1945, it has been the site of 16,000 launches, from orbital and suborbital rockets to high-altitude balloons and unmanned aerial vehicles.

Astrotourists looking for the traditional rocket launch experience at Wallops should keep an eye on launch manifests that mention a planned lift-off from MARS. The Mid-Atlantic Regional Spaceport (MARS), on the Wallops Flight Facility grounds, is where you're most likely to see a rocket launch to space. Launches typically occur from MARS about once a year, so plan ahead if you're travelling to the region specifically for a launch. You can also visit the Wallops Visitor Center to learn more about NASA research and launches conducted from the site.

It's a three-hour drive from Vandenberg to the NASA Jet Propulsion Laboratory in Pasadena, California (p134). Base yourself in Los Angeles, and you can make it a multi-day road trip through Southern California's astrotourism highlights. Also consider visiting Joshua Tree National Park, two hours east of LA, to see this otherworldly destination under the Milky Way.

Itinerary Extender

No trip to the Washington, DC region would be complete without a stop at the Smithsonian National Air and Space Museum. Visitors can explore dozens of exhibits on space exploration and human space flight. The retired Space Shuttle *Discovery* is also on display, and visitors can explore a full-scale replica of the shuttle's mid-deck.

Space Tourism

While humans have looked to the night sky for millennia, seeking answers and considering the meaning of the stars, planets and astronomical phenomena above, it's only relatively recently that we've begun our journey among the stars. First imagined by early science fiction writers such as Jules Verne, who wrote ***From the Earth to the Moon*** (1865), and later by HG Wells in ***The First Men in the Moon*** (1901), space travel was firmly established in speculative fiction territory just a century ago. The 'space opera' genre went on to become wildly popular in the start of the 20th century, as science seemed to put space in reach.

Fact eventually followed fiction. The first man-made object in space was the Russian satellite Sputnik, in 1957. By April 1961 the Russians had achieved another first, sending cosmonaut Yuri Gagarin to space. Gagarin was also the first person to complete an orbit around the planet. While astrotourists wait for access to the space market to mature, they can visit the launch site where the then-USSR sent Gagarin and his fellow cosmonauts off on their journey, Kazakhstan's Baikonur Cosmodrome (p258). The US sent its first astronaut, Alan Shepard, into space that same year, and astronaut John Glenn was the first American to orbit the Earth in 1962. The storied heroics and accomplishments of the Apollo astronauts in the 1960s and early 1970s helped secure American dominance in the space race, as Neil Armstrong and Buzz Aldrin became the first men to set foot on another celestial body, the moon, in 1969. In total, over 530 other people have been to space, the vast majority of them scientists, researchers and pilots.

Beginning in the 1970s, intrepid aerospace businesspeople began to dream about a world where everyday citizens could go to space. As the US space shuttle programme began, the designers envisioned the shuttle carrying a passenger cabin with up to 74 people to space. These plans were never fully realised, as disasters like the space shuttle ***Challenger*** (1986) and the space shuttle ***Columbia*** (2003) slowed the development of reusable shuttle launch technology. While some companies tried to create a market for space tourism in the 1990s and 2000s, technology and cost made it virtually impossible to capture or grow consumer interest.

In 2001 American Dennis Tito became the first space tourist when he booked a $20 million ticket to space using a space travel agency called Space Adventures. Tito had originally booked a ticket to the Russian space station Mir, but when that was decommissioned in 2001, his ticket was transferred to the International Space Station (ISS). Tito spent seven days aboard the ISS, helping with research and admiring the view. Since then, six other individuals from around the world have visited the ISS, each paying $20 to $40 million for the privilege.

The 2010s was the first decade when space tourism finally seemed like a possibility and a reality, though ticket prices are still astronomical by the standards of most travellers. Private industry is the primary driving force in the space tourism market. Major players have included corporations like XCOR, Virgin Galactic, the Sierra Nevada Corporation, Blue Origin (owned by Amazon founder Jeff Bezos) and Elon Musk's SpaceX, as well as less-known companies like Zero2Infinity, World View Enterprises, and Stratolaunch. Some of these companies are still operating; others have failed in the attempt to make space affordable and accessible. Even with ticket prices ranging from USD$75,000 to $250,000 depending on the company and the launch vehicle, the space tourism market is young and growing.

Earth-Based Space Tourism

ASTRONAUT TRAINING AND HABITATS

You may think you need to leave Earth to experience space tourism, but there are already ways to encounter space-like phenomena on our planet. Typically this involves doing activities similar to those astronauts do for training.

One common Earth-based space activity is reduced-gravity flights. In this experience, passengers board a modified airplane that proceeds to make a series of parabolic arcs, changing altitude at a 45° angle towards and away from the Earth. These arcs create the sensation of increased and decreased gravity; the decreased gravity can be made to simulate weightlessness similar to what is experienced on the International Space Station, gravity on the moon (16.6% of Earth) or gravity on Mars (38% of Earth). There is one public reduced-gravity tour provider, Zero G Corporation. It operates 'zero-g' flights around the United States.

Another activity is astronaut

From left: Space Camp in Huntsville, AL; a display of our solar system at the Cosmodome in Québec.

© Michael Doolittle / Alamy Stock Photo; © Mario B Beaustock / Alamy Stock Photo

training. Space Camp in the United States is the most well known (p136), but there other public 'training' facilities too, including the Cosmodome in Canada and Space Camp Russia in Star City, Russia. At these events participants can learn about the science and the Earth-based operations that go into manned space flight. You can also spin in a multi-axis trainer, move in simulated reduced-gravity chairs and work with equipment that reproduces different environments on the ISS. At Space Camp in Huntsville, Alabama, participants can also train in a swimming pool similar to the one astronauts train in at Johnson Space Center in Houston. Most astronaut training camps like this are open to participants of all ages, including children.

There are other interesting Earth-based space tourism activities including habitat simulations, where participants travel to a place that simulates the moon or Mars and live/ behave as though they are on another celestial body (see sidebar for more information on these). People can also experience virtual reality simulations, which have improved enough to give users the sense that they are in space or on another planet.

With space tourism options still limited in scope or prohibitively expensive for most, these Earth-bound simulations are a chance to imagine and dream about the possibilities of future human exploration of space.

Visitors to Biosphere 2, outside of Tucson Arizona, can learn about the biosphere experiments held here between 1991 and 1994. Eight participants were sealed inside the dome, which included examples of Earth's main climate zones. Intended to be self-supporting, a decline in oxygen over time meant it had to be introduced from outside. Nevertheless, it was still a bold effort probing the viability of enclosed ecological systems to support and maintain human life in outer space. NASA and the University of Hawai'i continue to run HI-SEAS, a similar programme focused on Mars. In a dome on Hawai'i's Mauna Loa, four participants are enclosed in an environment built as an analogue for the conditions in exploration scenarios.

Suborbital Space Tourism

ROCKETS, SPACEPLANES AND MORE

Most space tourism projects focus on so-called suborbital space tourism. Suborbital is used to denote that the trip will occur close to Earth, only 62–99 miles (100–160km) above the planet. While this sounds like a great distance, 62 miles (100km) is considered the Karmán Line, the line between Earth and space, meaning trips to this altitude are only barely entering space. Suborbital space tourism is a major opportunity for space tourism, as it allows passengers to see the Earth from extreme altitude and thus see the curve of the planet and the tiny shell of Earth's atmosphere protecting us. The idea of these flights is to reach an altitude at which passengers can experience weightlessness, whether for a few minutes or for hours.

There are three main ways companies are developing launch technology for suborbital space tourism: rockets, spaceplanes and balloons. Rocket-focused space tourism companies, like Blue Origin, create a rocket and passenger capsule that launch to space together, then separate and return to Earth. Blue Origin plans to reuse the first-stage booster of each New Glenn rocket, named after the pioneering astronaut John Glenn, for considerable cost savings. Passengers enjoy a weightless experience for a few minutes as the capsule begins to fall back down; parachutes and booster engines help the capsule land softly.

Plane-based suborbital flight is another option. Spaceplanes use the technology behind airplanes to help put the passenger capsule or rocket closer to space, making it easier and more affordable to send people to space. Companies like Virgin Galactic and Stratolaunch, in comparison to rocket-based launches, intend to use a plane to launch passengers to suborbital space. By affixing a spaceplane to the bottom of a more conventional airplane, these companies hope to launch from a higher altitude, reducing the cost and fuel needed to make the trip. Virgin Galactic's SpaceShipTwo reaches an altitude of 50 miles (80km) and aims higher,

Virgin Galactic workers gather by White Knight Two during the rollout of SpaceShipTwo.

© zuma Press, Inc. / Alamy Stock Photo

repeatedly breaking the supersonic sound barrier on their flights.

Lastly, balloons are an unusual but cost-effective way to send people to space. Companies like Zero2Infinity and World View Enterprises have borrowed from weather balloon technology to create large balloons that lift a passenger capsule to space. Imagine the highest sky-diving trip possible. When the balloon reaches a certain altitude, it bursts, and the passenger capsule begins its descent to Earth aided by parachutes. In each case the experience provides minutes or hours of weightlessness during the apex of their flight. Buyers should be aware that the ticket price for these suborbital flights is high compared to the duration of the experience.

Suborbital Space Tourism Companies & Ticket Prices:

Blue Origin
Around $250,000 per person

Virgin Galactic
$250,000 per person

Zero2Infinity
$143,000 per person

World View Enterprises
$75,000 per person

Has SpaceShipTwo actually reached space? It depends who you ask. Virgin Galactic say yes and US agencies do give out astronaut wings to pilots who go above a height of 80km. But the FAI, responsible for setting standards on international space travel, defines the border as the Kármán Line at 100km, meaning Virgin Galactic's craft was not tested in this environment.

Orbital Space Tourism

SPACE STATIONS

A mock-up of Orion Span's planned space hotel module.

Courtesy of Orion Span

Orbital space tourism goes beyond the altitude of suborbital space tourism and allows you to experience life while actually orbiting Earth. Space tourism opportunities at this altitude generally last between days and weeks; future endeavours may even allow space tourists to spend one or two months in space.

One notable option is visiting the International Space Station as a space tourist. This is a complicated process involving both a large ticket price – $55 million was the last amount quoted by the Russian space agency, Roscosmos – and months of training in advance of the trip. The main space tourism operator to the ISS is Space Adventures, which provides this trip and others to customers who are committed and able to pay.

Private, non-research space habitats are already in orbit and will continue to be developed in future years, despite the challenges. Bigelow Aerospace was the first company to send inflatable habitat modules to space; two such modules have been orbiting since 2006 and 2007, respectively. Bigelow hopes to build a 'space hotel' from these inflatable modules where visitors can stay for extended periods of time as they orbit. Other companies, including Orion Span, have proposed similar ideas. Orion Span hopes to launch their Aurora Space Station habitat for a 12-day trip at the cost of a cool $9.5 million on a vessel designed to include a hologram deck, viewing windows and even wi-fi, with room for two crew members and four space tourists on board. Axiom Space also hopes to launch its design-oriented space station and to welcome space tourists for $55 million for an eight-day trip to space. All three companies hope to be established in space by 2022.

What can you experience on an orbital space flight? In addition to a view of Earth from further away than suborbital flights allow, travellers would experience the darkness of space, have a chance to see the aurora dancing across Earth's atmosphere and watch the sunrise and sunset 16 times per day as their space hotel or space station orbits around the Earth!

Visions of space exploration often bring up the enclosed surface habitat in *The Martian* **or the paranoia-tinged world of sci-fi classic** *2001: A Space Odyssey*, **set on an interplanetary mission. Today's companies envision something closer to the world of** *Solaris*, **set on a space station in orbit around a distant moon, though without the pesky psychological delusions, of course.**

From the Moon to Mars

INTERPLANETARY PLANS

Beyond the moon, the next natural destination for space tourism is Mars. Both NASA and private space companies like SpaceX have set their sights on establishing a permanent settlement on Mars in the coming decades, which will open up new tourism opportunities. While we have only sent satellites, landers and rovers to Mars so far, we've learned a lot about what life might be like on the planet and how much we'll need to prepare to survive and thrive on Martian soil. It will likely be several decades before humans are able to establish a settlement on Mars and develop the infrastructure to support tourism. It's also a multi-month trip to reach Mars, so you'll need to save up some holiday time when you start planning your Martian vacation.

When we imagine the tourism opportunities on Mars, it's important to remember that most of the scientific data we have about Mars so far suggests it was once a planet quite similar to Earth. Many of the adventurous travel experiences we have on Earth will be even more dramatic on Mars. There are mountains to scale, including the massive volcanic Olympus Mons, which towers 85,000ft (25,908m) over the Martian plains. The Martian equator is dominated by the Valles Marineris, or Mariner Valley, which runs some 2500 miles (4025km) and is up to 4 miles (6.4km) deep in parts; it's ripe for exploration and rock climbing. And don't forget about the underground destinations to explore. Mars is dotted with caves and subterranean systems – some of which may hold water or ice that will be crucial to human settlement. Spelunking on Mars will be an opportunity to both explore and better understand the history of the planet.

The thin atmosphere on Mars rules out some adrenaline-inducing activities like skydiving and paragliding with current technology, but these obstacles will surely be overcome. Mars' gravity, which is only 38% of the gravity on Earth, also opens up new opportunities we can't quite imagine.

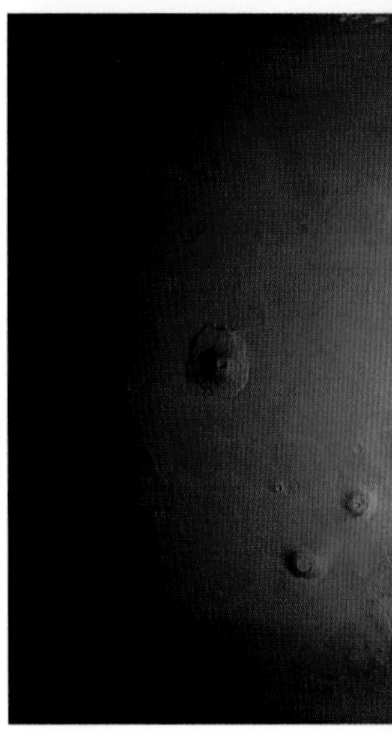

All images courtesy of Kevin M. Gill

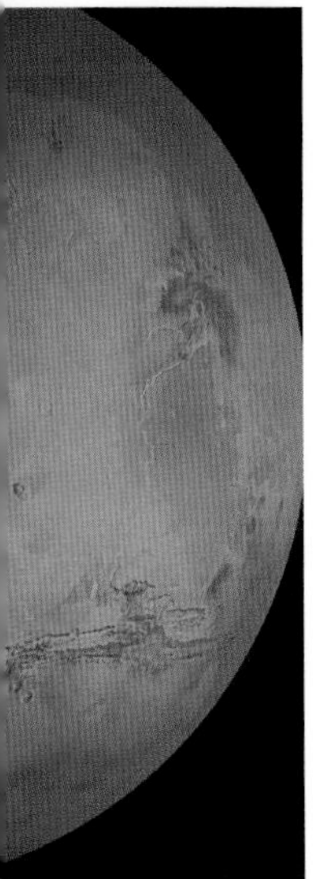

Clockwise from top: A rendering of a SpaceX Mars outpost; the surface of Mars; an image of the Red Planet as gathered from NASA HiRISE data.

'Follow the Water' has been one of the major goals of NASA's Mars programme. Water is currently driving NASA's exploration into the outer solar system, where ocean worlds such as certain moons of Jupiter and Saturn hold the potential to support life.

Data from the European Mars Express spacecraft, obtained by a radar instrument called MARSIS (Mars Advanced Radar for Subsurface and Ionosphere Sounding), indicates that there may be liquid water left on Mars. A 'bright spot' was detected in radar signals about 1 mile (about 1.5 km) below the ice cap in the Planum Australe region. This strong radar reflection was interpreted as liquid water – one of the most important ingredients for life in the universe.

Solar System and Interstellar Tourism

TO INFINITY AND BEYOND

Space tourism beyond Earth, the moon and Mars sounds exciting, but where might we go? NASA's Jet Propulsion Lab released 'travel posters' in 2018 to inspire dreams of travel through our solar system and beyond. Posters featured such destinations as Venus, the asteroid Ceres and Jupiter's moon Enceladus, as well as experiences like watching the aurora on Jupiter or taking a 'Grand Tour' following the same route as the spacecraft Voyager. These posters might dramatise or exaggerate exactly which kinds of space tourism you might get to experience, but they inspire us to think bigger about just how far we can go.

JPL designers also created travel posters for much further-afield

From left: ESO's La Silla Observatory; a rendering of NASA's Cassini on its mission to Saturn.

From left: © Science History Images / Alamy Stock Photo; © Kevin M Gill / Creative Commons

exoplanets, planets in orbit around other stars; new advances in detection abilities have begun discovering exoplanets en masse. They include the so-called 'super Earth' HD 40307 g, which is 42 light years away from Earth in the vicinity of the constellation Picton. This terrestrial planet is one of six known to orbit dwarf star HD 40307 and is believed to be within the habitable 'Goldilocks zone' of its star, close enough to receive warming rays but not too close as to get burned to a crisp. HD 40307 g is six times the mass of Earth; if humans were ever to visit the exoplanet, they would find a gravity much stronger than that on our home planet. This is just one of hundreds of exoplanets discovered by the European Space Observatory's ground-based planet-hunting instrument HARPS at La Silla in Chile's Atacama Desert (p114) and by the Kepler and TESS spacecraft.

While it's hard to imagine exactly where or how far our desire for intergalactic exploration might take us someday (beyond the Kuiper belt? Beyond the Milky Way?), there's no doubt that tourism opportunities will follow in the footsteps of the scientists and pilots who will lead the way. It might sound like science fiction, but Verne and Wells wrote their stories of men visiting the moon as science fiction, and travel to the moon was realised before a century passed. Near our lifetimes, the whole universe might well be open to us as explorers and travellers.

The latest plans for manned space exploration from NASA focus on Orion, a manned spacecraft meant to travel farther into the solar system than humanity has ever gone. The astronauts' mission will be to confirm all of the spacecraft's systems operate as designed in the actual environment of deep space with crew aboard. 'During this mission, we have a number of tests designed to demonstrate critical functions, including mission planning, system performance, crew interfaces, and navigation and guidance in deep space,' said Bill Hill, deputy associate administrator. 'It's just like the Mercury, Gemini, and Apollo programmes, which built up and demonstrated their capabilities over a series of missions.'

Conclusion

While this book might make you think you've got to jet around the world to experience the wonders of space and the night sky, the reality is that you can become a stargazing astrotourist right at home. Most cities have space museums, observatories, planetariums, astronomy clubs and pockets of dark sky where you can enjoy the stars without boarding a plane. You don't need to spend your entire vacation on astrotourism either; it's easy to add on a planetarium or museum visit to your daily activities, or plan for stargazing at night once your day of adventures is over. Take advantage of space and astrotourism when you travel by researching the options in advance and travelling with any specialised viewing gear you may want to use.

As the dark-sky movement grows and more locations work to reduce light pollution, even more options will exist for witnessing the beauty of the stars, planets, galaxies and more above. Similarly, look for increased travel opportunities and destinations focusing on space experiences. Some of the destinations mentioned in this book (like San Pedro de Atacama in Chile, and Merzouga and Erg Chebbi in Morocco) already focus on astronomy and stargazing as a main tourism draw. Others take advantage of space experiences to draw travellers, like the Space Coast in Florida; the eclipses across North America in 2024 and Europe in 2026 will also attract visitors. This kind of tourism is only going to increase in coming years, and travellers who want to add space experiences to their itinerary, or even focus their entire itinerary on astronomy, will have more and more choices.

Though it may seem far-fetched given current ticket prices that cause all but the ultra-wealthy a good case of sticker shock, space tourism will also likely become more affordable in the coming decade. The high costs of space travel today, as with most new destinations and travel experiences, will be reduced through new technology, greater access and more tourism. Antarctica once seemed out of reach but is now a once-in-a-lifetime trip for an increasing number of travellers; in the same way, someday you may board a rocket headed into low Earth orbit, the moon or even Mars.

No matter how you choose to experience space in your everyday life, it's important to keep it all in perspective. One of the most powerful impacts of seeing the night sky is better understanding how precious and special our home planet is in this great big universe. While the planet is a big place full of many wonders, it's also our only home (for now!), and we need to take care of it. Don't be surprised if you find yourself more conscientious about environmental issues or resource management after you start experiencing the wonders of space first-hand. This is a common reaction to viewing the Milky Way, standing in awe of the aurora or being humbled during a total solar eclipse. If astrotourism helps more people protect our amazing home planet, the future will be bright ... and the night skies will be dark and full of stars.

Index

Dark Skies
September 2019
Published by Lonely Planet Global Limited
CRN 554153
www.lonelyplanet.com
10 9 8 7 6 5 4 3 2 1
Printed in Singapore
ISBN 978 1 78868 6198

Managing Director, Publishing Piers Pickard
Associate Publisher Robin Barton
Written by Valerie Stimac
Commissioning Editor Nora Rawn
Art Director Daniel Di Paolo, Katharine Van Itallie
Print Production Nigel Longuet

Lonely Planet Offices

Australia
The Malt Store, Level 3,
551 Swanston St, Carlton, Victoria 3053
T: 03 8379 8000

USA
124 Linden St, Oakland,
CA 94607
T: 510 250 6400

Ireland
Digital Depot, Roe Lane (Off Thomas Street)
The Digital Hub,
Dublin 8, D08 TCV4

Europe
240 Blackfriars Rd,
London SE1 8NW
T: 020 3771 5100

STAY IN TOUCH lonelyplanet.com/contact

COVER IMAGE © Paul Zizka / Aurora Photos

Paper in this book is certified against the Forest Stewardship Council™ standards. FSC™ promotes environmentally responsible, socially beneficial and economically viable management of the world's forests.